I. Pichlmayr · U. Lips · H. Künkel

The Electroencephalogram in Anesthesia

Fundamentals, Practical Applications, Examples

Translated by E. Bonatz and T. Masyk-Iversen

With 70 Figures

Springer-Verlag
Berlin Heidelberg New York Tokyo 1984

Professor Dr. INA PICHLMAYR
Privat-Dozent Dr. ULRICH LIPS
Zentrum für Anästhesiologie der Medizinischen Hochschule Hannover, Abt. IV
Krankenhaus Oststadt, Podbielskistr. 380, D-3000 Hannover 51

Professor Dr. HELMUT KÜNKEL
Zentrum Neurologische Medizin, Abt. II – Institut für Klinische Neurophysiologie und Experimentelle Neurologie, Konstanty-Gutschow-Str. 8, D-3000 Hannover 61

Translators:
E. BONATZ
T. MASYK-IVERSEN
Zentrum für Anästhesiologie der Medizinischen Hochschule Hannover, Abt. IV
Krankenhaus Oststadt, Podbielskistr. 380, D-3000 Hannover 51

Title of the original German edition 1983:
Das Elektroenzephalogramm in der Anästhesie

ISBN-13: 978-3-642-69564-3 e-ISBN-13: 978-3-642-69562-9
DOI: 10.1007/978-3-642-69562-9

Library of Congress Cataloging in Publication Data. Pichlmayr, I. (Ina), 1932–. The electroencephalogram in anesthesia. Translation of: Das Elektroenzephalogramm in der Anästhesie. Includes bibliographies and index. 1. Anesthesiology. 2. Encephalography. 3. Patient monitoring. I. Lips, U. (Ulrich), 1948–. II. Künkel, Helmut. III. Title. [DNLM: 1. Anesthesia. 2. Electroencephalography. 3. Monitoring, Physiologic methods. WL 150 P592e] RD82.P513 1984 617'.96 84-5569

Softcover reprint of the hardcover 1st edition 1984

2119/3130-543210

Preface

Neurological, neurovegetative, and psychic changes following surgery or intensive care indicate alterations in cerebral function. When these changes are analyzed exactly, they are found to be by no means rare; at the same time, they present a wide spectrum of clinical manifestations. While optimal circulatory monitoring – including continuous EEG recording during anesthesia or intensive care – is today regarded as routine, the problem of objective and continuous monitoring of cerebral function under various forms of anesthesia, surgery, and intensive care has not been solved.

Considerable improvements in clinical neurophysiology, mainly through visual and spectral analytic evaluations of the EEG, show functional neurophysiological changes to be equivalent to, or the cause of, clinical, neurological, and psychic changes. The question of current interest, therefore, is whether anesthesiological EEG monitoring is able, both theoretically and practically, to detect and adequately interpret cerebral alterations following perioperative measures such that these alterations can be treated or avoided accordingly. In the long run, the basic questions are to what extent can anesthesiological measures be guided by the EEG pattern and managed effectively, and can EEG analysis be significant both for the individual patient and in the general interpretation of the effect of anesthesiological procedures?

Questions of the relevance and practicability of anesthesiological EEG monitoring are dealt with in this book. This study is based on an analysis of more than 1500 EEGs of patients who underwent anesthesia during surgery. The effects of premedication, anesthesia, events during the perioperative period, and intensive care are analyzed separately. Theoretical and practical aspects of anesthesiological EEG monitoring are outlined in the Introduction. Results and practical experience from all sources give a perspective on the routine use of EEG monitoring in the future.

We would like to thank the Volkswagenwerk Foundation for their generous support of the clinical studies. We are grateful to our assistants Mrs. U. Lessing, Mrs. K. Redeker, Mrs. H. Schulze, and Mrs. K. Schwedhelm.

Ina Pichlmayr, Ulrich Lips, Helmut Künkel

Contents

Detailed contents are found at the beginning of each chapter

A. Introduction to Electroencephalography in Anesthesiology

I. Development and Application of Electroencephalography

The electroencephalogram is a neurophysiological technique for examining cerebral electric activity; it can be applied to many medical fields and has allowed considerable advances to be made. This has been due to the development of data processing and signal analysis as well as to experience in the clinical application of the findings (JUNG 1953; COPPER et al. 1974; DOLCE and KÜNKEL 1975; REMOND 1977). Berger, whose name is linked inseparably with the history of electroencephalography, hoped to determine the mechanisms of psychic processes with this new harmless method (KOLLE 1965). His expectations were not fulfilled, but he did recognize the value of electroencephalography as an experimental method in the study of drug-induced changes of consciousness. In 1933, he performed the first electroencephalogram (EEG) recording under chloroform anesthesia. Subsequently, barbiturates, synthesized in 1932, were thoroughly studied with the EEG [27]. This led to the development of EEG-controlled dosage machines [23]. Research in clinical neurophysiology with respect to the technique and validity of electroencephalography revealed further possibilities for application. In anesthesiology, the EEG has proven useful in special indications (e.g., open heart surgery, intra- and postoperative cerebral emergency). Monographs about the application and value of the EEG, especially for the anesthesiologist, were published by BRECHNER et al. (1962), SADOVE et al. (1967), PRIOR (1979), and GRABOW (1981). In the light of the, on the whole, promising advances that have been made, the desirable routine EEG monitoring of anesthesiological measures would appear to be practicable in the near future. Milestones in the historical development of electroencephalography, briefly summarized and without claiming completeness, are given below, outlining the development to date.

1875 R. CATON (London) [30]
First tracing of cerebral electric potentials in animal experiments by galvanometry with optical amplification.

1890–1893 B. DANILEWSKI (Charkow) [34], F. VON MARXOW (Vienna) [64, 65], A. BECK, N. M. CYBULSKI (Krakau) [4, 5]
Tracings of electric processes in various animal species with inconsistent results, due to poor technical conditions. In 1890, von Marxhow studied the effect of chloroform anesthesia on the animal EEG.

1904 S. Tschiriev (Kiev) [78]
Hypothesis about the origin of electric potentials through variable cerebral congestion.

1912 P. I. Kaufmann (St. Petersburg) [56]
Final refutation of Tschiriev's theory.

1913–1925 N. W. Prawdicz-Neminski (Kiev) [66, 67]
Experimental elaboration on recordings of the EEG background activity in animals; his results are still valid today.

1914 N. M. Cybulski (Krakau) [33]
Description of EEG recordings during epileptic seizures brought about by electric stimulation of the canine cortex.

1924 H. Berger (Jena) (Brazier 1961)
First EEG recording on a human subject performed using silver-needle electrodes. Discovery of the alpha activity over the occipital region.

1929 H. Berger (Jena) [7]
First publication on the EEG in man.

1930–1938 H. Berger (Jena) [8–21]
The results of 13 further papers and one publication summarizing the fundamentals, criteria for assessment, and possibilities for development of the EEG; some of the results are still valid today.

1932 G. Dietsch, H. Berger (Jena) [35]
First application of the Fourier analysis in short EEG recordings.

1932–1933 J. F. Tönnies (Berlin) [74–76]
Development of the "neurograph". This EEG apparatus with continuous registration by means of pen replaced the complicated photographic documentation that had previously been used.

1933 H. Berger (Jena) [12]
First EEG recordings during chloroform anesthesia in man.

1934 E. D. Adrian, A. Matthews (Cambridge) [1]
Acknowledgement of Berger's discoveries and scientific studies, which led to the final recognition of Berger's work in 1937 during the psychology congress in Paris. Improvement of the EEG recording technique by means of a Faraday cage.

1932–1937 A. E. Kornmüller (Berlin) [58, 59]
Discovery of various activities in the cortical fields in animal experiments.

1934–1935 J. F. Tönnies, D. Förster, H. Altenburger (Berlin) [44, 76]
First EEG recordings from the cortex during craniocerebral surgery.

1934–1937 H. Rohracher (Vienna) [69–71]
Postulation of the so-called alpha-wave theory for explaining psychiatric disorders.

1935 F. A. Gibbs, H. Davies, W. G. Lennox (Boston) [45]
First description of spike-wave patterns in the EEG of patients with clinical signs of petit mal epilepsy.

1936 W. G. Walter (Bristol) [81, 82]
Demonstration of parameters for tumor localization in the brain.

1937 F. A. Gibbs, E. L. Gibbs, W. G. Lennox (Boston) [46]
First extensive study on the correlation of anesthesia and EEG changes.

1938 H. A. Grass, F. A. Gibbs (Boston) [50]
Elaboration of a method for applying the Fourier transformation to the EEG.

1938 Z. Drohocki, J. Drohocka (Brussels) [37]
Animal experiments revealing severe impairment by anesthesia of the cortical area of the brain as opposed to deeper cerebral areas.

1939 Z. Drohocki (Brussels) [36]
Quantification of the EEG by means of a graphic amplitude integration.

1943–1946 G. R. Baldock, W. C. Walter (Bristol) [3]
Extension of frequency analysis of the EEG by use of electronic filters; the further development of this method was hindered initially by problems with electronic technology.

1945 M. A. B. Brazier, J. E. Finesinger (Los Angeles) [27]
Studies on the effect of barbiturates on the EEG.

1950 R. F. Courtin, R. G. Bickford, A. Faulconer (Rochester) [32]
Findings indicating a correlation between EEG changes and depth of anesthesia using combined nitrous oxide/oxygen/ether anesthesia.

1949–1951 R. G. Bickford (Rochester) [22–24]
Development of a device for automatic control of anesthesia, based on EEG changes determined in animal experiments and occasionally used clinically.

1951 D. K. Kiersey, R. G. Bickford, A. Faulconer (Rochester) [57]
Definition of stages of anesthesia by means of specific EEG changes.

1951–1952 W. G. Walter and H. W. Shipton (Bristol) [83], A. Remond and F. Offner (Paris) [68]
Development of the "toposcope", a device for optically controlled topographical recording of EEG frequency distribution.

1952 J. SCHNEIDER et al. (Colmar) [73]
Further differentiation of the classification of stages of anesthesia according to EEG pattern, today still valid for bartiturates.

1952 M. A. B. BRAZIER, J. U. CASBY (Los Angeles) [28]
First statistical processing of EEG parameters by auto- and cross-correlation functions.

1957/1959 B. SALTZBERG, A. R. BURCH (New Orleans) [29, 72]
Use of periodic EEG analysis. After initial technical problems, this method has become established especially in psychopharmacology by Fink and Itil.

1959 J. T. MARTIN, A. FAULCONER, R. G. BICKFORD (Rochester) [63]
Summary of the correlation between anesthesia and EEG changes, based on studies between 1933 and 1959.

1959 M. FINK, B. BENTE, T. M. ITIL (New York) [6, 39]
First systematic listing of EEG effects after application of phenothiazines. The basic work was carried out independently by the different authors.

1959–1960 ST. KUBICKI, M. TREDE, O. JUST (Berlin) [60]
Use of EEG monitoring in open heart surgery with extracorporal circulation.

Since 1961 T. M. ITIL (New York) [53–55]
EEG classification of psychopharmacological agents, first the group of phenothiazines, then the psychotropic drugs. Periodical EEG analyses are used for quantification of the EEG data.

1961–1969 M. FINK, P. BORENSTEIN et al.; L. GOLDSTEIN et al. (Massachusetts) [25, 26, 40–43, 48, 49]
Development of various EEG classifications for psychotropic drugs on the basis of different qualitative and quantitative properties.

1963 D. O. WALTER (Los Angeles) [80]
Further development of EEG spectral analysis.

1965 J. W. COOLEY, J. W. TUKEY [31]
Introduction of the fast Fourier transform with a reduction in the calculating effort required for Fourier analysis.

1967 G. DUMERMUTH, H. FLÜHLER (Zurich) [38]
Use of the fast Fourier transform (FFT) for the spectral analysis of the EEG. With the introduction of better computers and the reduction in the effort required to calculate the FFT, wide application of EEG spectral analysis becomes possible.

1967 M. S. SADOVE, D. BECKA, F. A. GIBBS (Rochester) (SADOVE et al. 1967)
Publication of the applications of the EEG in surgery. Despite technical difficulties, the application in anesthesiology and open heart surgery is strongly suggested.

1968–1974 T. M. Itil (New York) [54, 55]
Using analogous and digital computer methods for assessing the EEG, an extensive classification of psychotropic substances is established on a quantitative basis.

1970–1973 B. Hjorth (Solna) [51, 52]
Development of a new quantitative analytic method limiting the EEG information to three parameters. Despite the reduction in data, this method has advantages in that simultaneous application of the FFT to several EEG tracks becomes possible.

1972 H. Künkel (Hanover) [61, 62]
Description of a method with simultaneous application of the FFT to several EEG channels (up to 16) in real time. Thus, Fourier analysis becomes significant for topographical problems.

1974 L. Goldstein (Massachusetts) [47]
Re-establishment of Drohock's amplitude-integration method. As a result of technical improvements, it can be used in the study of psychotropic substances.

1976 G. Ahlbom, L. H. Zetterberg (Stockholm) [2]
First comparative studies of the already established various methods of frequency analysis.

1978 G. A. Volgyesi (Toronto) [79]
Introduction of a simple one-channel EEG analytic device for anesthesiology. Analysis is based on the calculation of the delta/alpha ratio.

Since 1975 Application of EEG techniques with and without frequency analysis in research, mainly in psychiatry and psycho-pharmaco-electroencephalography; in clinical fields in the specialties of neurology, psychiatry, internal medicine, heart surgery [77], intensive care, and anesthesiology.

Electroencephalography Meetings

1. International Meetings

1947 First international congress of electroencephalography and clinical neurophysiology (London), with conferences every 4 years since 1953.

2. European Meetings

1976 First European congress of electroencephalography and clinical neurophysiology (Venice); congress every 3 years.

3. National Meetings

As representative of the numerous meetings of the various national EEG societies, the yearly conference of the "Deutsche EEG-Gesellschaft" will be mentioned here.

1950 First Meeting (Wiesbaden)

1947 International EEG federation established (International Federation of Societies for Electroencephalography and Clinical Neurophysiology). Plans for the founding of an EEG society during the first international EEG meeting in London; actual foundation 1949 in Paris.

General Secretary – R. J. Ellington, 602 South 44 Avenue, Omaha, NE, 68105, USA.

The German section of the international EEG federation has been represented by the Deutsche EEG-Gesellschaft since 1950. This society directly originated from the "EEG-Arbeitsgemeinschaft", which was founded in Wiesbaden.

Secretary – S. Kubicki, Krankenhaus Westend, Spandauer Damm 130, 1000 Berlin 19, Federal Republic of Germany.

Publications of conference reports since 1970 have appeared in the journals "EEG-EMG" and "EEG Journal".

Journals

Journal of Electroencephalography and Clinical Neurophysiology (EEG Journal) Editors: M. A. B. Brazier (Los Angeles), P. Buser (Paris). Publisher: Elsevier, Amsterdam. This journal is the official organ of the international EEG federation.

Zeitschrift für Elektroencephalographie, Elektromyographie und verwandte Gebiete (EEG-EMG). Editors: H. Caspers (Münster), R. Hess (Zurich), J. Kugler (Munich), H. Petsche (Vienna), A. Struppler (Munich). General editors: S. Kubicki (Berlin). Publisher: Thieme, Stuttgart. This journal is the official organ of the German EEG Society.

References

A. Textbooks and Summaries

Brazier MAB (1961) A history of the electrical activity of the brain. Pitman, London

Brechner VL, Walter RD, Dillon JB (1962) Practical electroencephalography for the anesthesiologist. Thomas, Springfield

Copper R, Osselton JW, Shaw JC (1974) Elektroenzephalographie. Fischer, Stuttgart

Dolce G, Kühnel H (1975) CEAN – computerized EEG-analysis. Fischer, Stuttgart

Grabow L (1981) Hirnfunktionen unter dem Einfluß der allgemeinen Anästhesie. Fischer, Stuttgart

Jung R (1953) Neurophysiologische Untersuchungsmethoden, II. Das Elektroenzephalogramm. Handbuch der inneren Medizin V/1. Springer, Berlin Göttingen Heidelberg, pp 1216–1325

Kolle K (1965) Große Nervenärzte, vols I–V. Thieme, Stuttgart

Prior PF (1979) Monitoring cerebral function. Elsevier, Amsterdam

Remond A (1977) EEG Informatics: A didacted review of methods and applications of EEG. Elsevier, Amsterdam

Sadove MS, Becka D, Gibbs FA (1967) Electroencephalography for anesthesiologists and surgeons. Pitman, London

B. Papers

1. Adrian ED, Matthews BHC (1934) The Berger-rhythm – Potential changes from the occipital lobes in man. Brain 57:355–385
2. Ahlbom G, Zetterberg LH (1976) A comparative study of five methods for analysis of EEG (Tech. Rept. No. 112). Royal Institute of Technology, Stockholm
3. Baldock GR, Walter WG (1946) A new electronic analyzer. Electron Eng 18:339–344
4. Beck A (1890) Die Bestimmung der Lokalisation der Gehirn- und Rückenmarksfunktion vermittels der elektrischen Erscheinungen. Zentralbl Physiol 4:473–476
5. Beck A, Cybulski NN (1892) Weitere Untersuchungen über die elektrischen Erscheinungen in der Hirnrinde der Affen und Hunde. Zentralbl Physiol 6:1–6
6. Bente D, Itil TM (1959) Clinico-electroencephalographic investigations. In: Psychopharmacology Frontiers. Little Brown, Boston, pp 319–323
7. Berger H (1929) Über das Elektroenkepalogramm des Menschen. Arch Psychiatr Nervenkr 87:527–570
8. Berger H (1930) Über das Elektroenkephalogramm des Menschen: II. Mitteilung. J Psychol Neurol (Leipz) 40:160–179
9. Berger H (1931) Über das Elektroenkephalogramm des Menschen: III. Mitteilung. Arch Psychiatr Nervenkr 94:16–60
10. Berger H (1932) Über das Elektroenkephalogramm des Menschen: IV. Mitteilung. Arch Psychiatr Nervenkr 97:6–26
11. Berger H (1932) Über das Elektroenkephalogramm des Menschen: V. Mitteilung. Arch Psychiatr Nervenkr 98:231–254
12. Berger H (1933) Über das Elektroenkephalogramm des Menschen: VI. Mitteilung. Arch Psychiatr Nervenkr 99:555–574
13. Berger H (1933) Über das Elektroenkephalogramm des Menschen: VII. Mitteilung. Arch Psychiatr Nervenkr 100:301–321
14. Berger H (1933) Über das Elektroenkephalogramm des Menschen: VIII. Mitteilung. Arch Psychiatr Nervenkr 101:452–469
15. Berger H (1934) Über das Elektroenkephalogramm des Menschen: IX. Mitteilung. Arch Psychiatr Nervenkr 102:538–557
16. Berger H (1935) Über das Elektroenkephalogramm des Menschen: X. Mitteilung. Arch Psychiatr Nervenkr 103:444–454
17. Berger H (1936) Über das Elektroenkephalogramm des Menschen: XI. Mitteilung. Arch Psychiatr Nervenkr 104:678–689
18. Berger H (1937) Über das Elektroenkephalogramm des Menschen: XII. Mitteilung. Arch Psychiatr Nervenkr 106:165–187
19. Berger H (1937) Über das Elektroenkephalogramm des Menschen: XIII. Mitteilung. Arch Psychiatr Nervenkr 106:577–584
20. Berger H (1938) Über das Elektroenkephalogramm des Menschen: XIV. Mitteilung. Arch Psychiatr Nervenkr 108:407–431
21. Berger H (1938) Das Elektroenkephalogramm des Menschen. Nova Acta Leopold 6:173–304
22. Bickford RG (1949) Neurophysiological applications of automatic anesthesia-regulator controlled by brain potentials. J Physiol 159:562–563
23. Bickford RG (1950) Automatic electroencephalographic control of general anesthesia. Electroencephalogr Clin Neurophysiol 2:93–96

24. Bickford RG (1951) Use of frequency discrimination in the automatic EEG-control of anesthesia. Electroencephalogr Clin Neurophysiol 3:81–85
25. Borenstein P, Cujo P, Chiua M (1965) A propos de la classification des substances psychotropes selon leurs effects sur l'électroencéphalogramme. Ann Med Psychol (Paris) 2:429–452
26. Borenstein P, Cujo P, Kramarz P, Champion C (1969) A propos de certains aspects électroencéphalographiques de l'action des psychotropes. Sem Hop Paris 45:1331–1336
27. Brazier MAB, Finesinger JE (1945) Action of barbiturates on cerebral cortex: Electroencephalographic studies. Arch Neurol Psychiatry 53:51–58
28. Brazier MAB, Casby JU (1956) Some applications of correlation analysis to clinical problems in electroencephalography. Electroencephalogr Clin Neurophysiol 8:325–331
29. Burch NR (1959) Automatic analysis of the electroencephalogram. Electroencephalogr Clin Neurophysiol 11:827–834
30. Caton R (1875) The electrical currents of the brain. Br Med J 2:278
31. Cooley JW, Tukey JW (1965) An algorithm for the machine calculation of complex fourier-series. Math Comp 19:267–301
32. Courtin RF, Bickford RG, Faulconer A jr (1950) The classification and significance of electroencephalographic patterns produced by nitrous oxide-ether-anesthesia during surgical operations. Proc Staff Meet Mayo Clin 25:197–206
33. Cybulski NN, Macieszyna J (1914) Prady cynnosciowe kory mózgowej (Action currents of the cerebral cortex). Bull Int Acad Cracovie Series B:776–781
34. Danilewski VI (1891) Zur Frage über die elektrischen Vorgänge im Gehirn als Ausdruck seines Tätigkeitszustandes. Zentralbl Physiol 5:1–4
35. Dietsch G, Berger H (1932) Fourier-Analyse von Elektroenkephalogrammen des Menschen. Pflügers Arch 230:106–112
36. Drohocki Z (1939) Elektrospectrographie des Gehirns. Klin Wochenschr 18:536–538
37. Drohocki Z, Drohocka J (1938) L'électrocorticogramme pendant l'establissement de la narcose à l'urethane. C R Soc Biol (Paris) 129:895–898
38. Dumermuth G, Flühler H (1967) Some modern aspects in numerical spectrum analysis of multichannel EEG-data. Med Biol Eng Comput 5:319–331
39. Fink M (1959) EEG and behavioral effects of psychopharmacologic agents. Neuropsychopharmacology 1:441–446
40. Fink M (1961) Quantitative electroencephalography and human psychopharmacology: Frequency spectra and drug action. Med Exp 5:364–369
41. Fink M (1963) Quantitative electroencephalography in human psychopharmacology. In: EEG and behavior. Glaser, New York, pp 177–197
42. Fink M (1969) The human electroencephalogram: Index of clinical activity of new psychoactive agents. Mod Probl Pharmacopsychiatry 2:106–110
43. Fink M (1969) EEG-classification of psychoactive drugs in man: Review and theory of behavioral associations. Psychopharmacology: A review of Progress 1957. US Government Printing Office, Washington, pp 497–507
44. Förster O, Altenburger H (1935) Elektrobiologische Vorgänge an der menschlichen Hirnrinde. Dtsch Z Nervenkr 135:277–288
45. Gibbs FA, Davis H, Lennox WG (1935) The electroencephalogram in epilepsy and in conditions of impaired consciousness. Arch Neurol Psychiatr 34:1135–1148
46. Gibbs FA, Gibbs EL, Lennox WG (1937) Effects on electroencephalogram of certain drugs which influence nervous activity. Arch Intern Med 60:154–166
47. Goldstein L (1974) Psychotropic drug induced EEG-changes as revealed by the amplitude integration method. Mod Probl Pharmacopsychiatry 8:131–148
48. Goldstein L, Murphree HB, Pfeiffer CC (1963) Quantitative electroencephalography in man as a measure of CNS stimulation. Ann NY Acad Sci 107:1045–1056
49. Goldstein L, Murphree HB, Sugerman AA, Pfeiffer CC, Jenney EH (1963) Quantitative electroencephalographic analysis of naturally occurring and drug induced psychotic states in human male. Clin Pharmacol Ther 4:10–21
50. Grass MA, Gibbs FA (1938) Fourier-transform of the EEG. J Neurophysiol (Springfield) 1:521–526
51. Hjorth B (1970) EEG-analysis based on time domain properties. Electroencephalogr Clin Neurophysiol 29:306–310

52. Hjorth B (1973) The physicial significance of time domain descriptions in EEG-analysis. Electroencephalogr Clin Neurophysiol 34:321–325
53. Itil TM (1961) Elektroenzephalographische Befunde zur Klassifikation neuro- und thymoleptischer Medikamente. Med Exp 5:347–363
54. Itil TM (1972) Quantitative pharmaco-electroencephalography in the discovery of a new group of psychotropic drugs. Dis Nerv Syst 33:557–559
55. Itil TM (1978) Quantitative electroencephalography in the discovery of psychotropic properties of drugs. Neuropsychopharmacology 2:1135–1140, 1183–1190
56. Kaufmann PI (1912) Elektrische Erscheinungen in der Großhirnrinde. Obrozenie Psikhiatrie Neurologii Eksperimenentalnoi Psikhiologii 17:403–514
57. Kiersey DK, Bickford RG, Faulconer A Jr (1951) Electroencephalographic patterns produced by thiopental-sodium during surgical operations: Description and classification. Br J Anaesth 23:141–152
58. Kornmüller AE (1932) Bioelektrische Charakteristika architektonischer Felder der Großhirnrinde. Psychiatr Neurol Wochenschr (Halle) 34:25–26
59. Kornmüller AE (1933) Die Ableitung bioelektrischer Effekte architektonischer Rindenfelder vom uneröffneten Schädel. J Psychiatr Neurol 45:172–182
60. Kubicki St, Trede M, Just O (1960) Die Bedeutung des EEG bei Herzoperationen in Hypothermie und bei extrakorporaler Zirkulation. Anaesthesist 9:119–123
61. Künkel H (1972) Die Spektraldarstellung des EEG. EEG EMG 3:15–24
62. Künkel H (1972) Simultane Viel-Kanal-On-Line-EEG-Analyse in Echtzeit. EEG EMG 3:29
63. Martin JT, Faulconer A Jr, Bickford RG (1959) Electroencephalography in anesthesiology. Anesthesiology 20:359–376
64. Marxow von F (1890) Mitteilung betreffend die Physiologie der Hirnrinde. Zentralbl Physiol 4:537–540
65. Marxow von F (1893) Mitteilung betreffend die Physiologie der Hirnrinde. Gesammelte Abhandlungen. Barth, Leipzig, S 409–411
66. Prawdicz-Neminski NW (1913) Elektrische Gehirnerscheinungen. Zentralbl Physiol 18:951–960
67. Prawdicz-Neminski NW (1925) Zur Kenntnis der elektrischen und der Innervationsvorgänge in den funktionellen Elementen von Geweben des tierischen Organismus. Elektrocerebrogramm der Säugetiere. Pflügers Arch Ges Physiol 209:362–382
68. Remond A, Offner F (1952) Etudes topographiques de l'activité EEG de la région occipitale. Rev Neurol (Paris) 87:182–189
69. Rohracher H (1935) Die gehirnelektrischen Erscheinungen bei geistiger Arbeit. Z Psychol 136:308–324
70. Rohracher H (1937) Die gehirnelektrischen Erscheinungen bei Sinnesreizen. Z Psychol 140:274–308
71. Rohracher H (1938) Experimentale und theoretische Untersuchungen über die gehirnelektrischen Vorgänge. Pont Acad Sci Comm 2:225–273
72. Saltzberg B (1957) A new approach to signal analysis in electroencephalography. IRE Trans Med Electron 8:24–30
73. Schneider J, Woringer E, Thomalske G, Brogly G (1952) Bases électrophysiologiques des mechanismes d'action du Pentothal chez le chat. Rev Neurol (Paris) 87:433–451
74. Tönnies JF (1932) Der „Neurograph", ein Apparat zur Aufzeichnung bioelektrischer Vorgänge unter Ausschaltung der photographischen Kurvendarstellung. Naturwissenschaften 20:381–384
75. Tönnies JF (1933) Der Neurograph, ein Apparat zur unmittelbar sichtbaren Registrierung bioelektrischer Erscheinungen. Dtsch Z Nervenheilk 130:60–67
76. Tönnies JF (1934) Die unipolare Ableitung elektrischer Spannungen vom menschlichen Gehirn. Naturwissenschaften 22/24:411–414
77. Trede M, Kubicki St, Just O (1959) Über EEG-Beobachtungen bei Herzoperationen mit extrakorporalem Kreislauf. Anaesthesist 8:76–82
78. Tschiriev S (1904) Propriétés électromotrices du cerveau et du coeur. J Physiol Pathogen 6:671–682
79. Volgyesi GA (1978) A brain function monitor for use during anesthesia. Preliminary report. Can Anaesth Soc J 25:427–430

80. Walter DO (1963) Spectral analysis for electroencephalograms. Mathematical determination of neurophysiological relationship from records of limited duration. Exp Neurol 8:155–181
81. Walter WG (1936) The location of cerebral tumors by electroencephalography. Lancet 2:305–308
82. Walter WG (1943) An automatic low frequency analyzer. Electron Eng 16:8–13
83. Walter WG, Shipton HW (1951) A new toposcopic display system. Electroencephalogr Clin Neurophysiol 3:281–292

II. Fundamentals of Electroencephalographic Analysis

CONTENTS

Electroencephalographic analysis using computers has been employed to a limited extent in clinical diagnostics [9, 14, 22]. It has been firmly established in sleep research and in the analysis of cerebral drug action. In anesthesiology and in monitoring during cardiac surgery (PRONK 1982), it has become increasingly significant. Here, meaningful use and critical assessment of EEG analysis require an understanding of the basic principles and possible interpretations of various analytic procedures.

What follows is a survey of some of the methods of quantitative EEG analysis. Basically, one can distinguish between techniques analyzing a certain *period of time* and methods analyzing a certain *frequency range*. For some analytic procedures, such a rather didactic and historical distinction is somewhat arbitrary; to a certain degree, the essence of methods using both time and frequency ranges may be comparable. A more thorough mathematical description will be omitted here as it is treated in other works with further literature references (PRONK 1982) [5, 7]. Furthermore, only those methods will be discussed that deal with the analysis of continuous basic activity relevant for anesthesiological problems. Methods of pattern recognition, e.g., for automatic recognition of a so-called epileptic activity, are described in the reviews of GOTMAN and GLOOR [13], JACOB [18], LOPES DA SILVA et al. [21], MCGILLIVRAY [23], and FERBER [8].

Methods of Time Analysis

Amplitude Integration. A comprehensive description of this method with its various modifications and a number of practical applications, especially for EEG in psychiatry, is found in GOLDSTEIN [12]. In the most commonly used version today, frequencies of below 0.75 Hz are eliminated from the signal by high-pass filtering; this reduces the practical problem of artifacts to a certain degree. Then there is a two-way rectification of the EEG signal with subsequent integration at selectable periods of time. An analysis performed in this way for each integration interval results in a figure which corresponds with the area beneath the rectified EEG pattern. This is illustrated in Fig. 1, the final reading of track 3 representing the integration value. Thus, the continuous EEG pattern produces a corresponding sequence of numbers that can be described by simple statistical values, such as mean, deviation, variation coefficient (deviation/mean ratio) or by distribution histograms. This can best be illustrated by an example. Figure 2 depicts the correlation between variation coefficient and integration values (mean energy content) for three sample groups. It can be concluded that the EEG in the dynamic structure of the basic activity, evidently depending on the patient's age, does contain sex-specific features, which remain hidden in visual EEG interpretation. The method of amplitude integration uses only a small fraction of the information contained in the EEG, but has the advantage of being technically easy to apply. In addition, its results can be interpreted very clearly and can be processed using simple statistical methods.

Period (Interval) Analysis. Basically; this method utilizes the idea of measuring intervals between points of time at which the EEG signal crosses

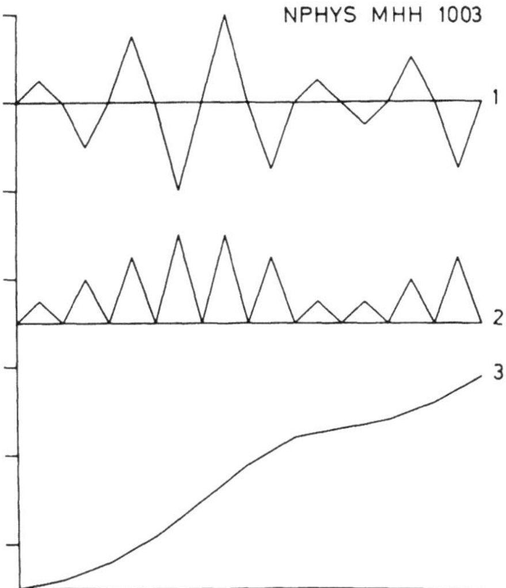

Fig. 1. The principle of amplitude integration. *1*, EEG pattern; *2*, rectified pattern; *3*, course of the integration value in the integration interval (after DROHOCKI [6])

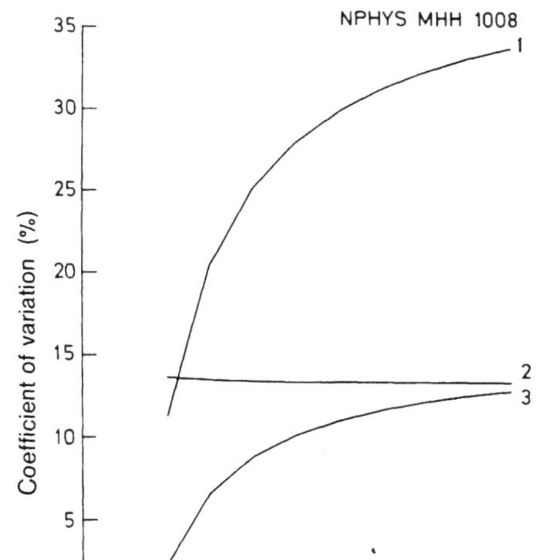

Fig. 2. Relation between the average of the "amplitudes" and the variation coefficient (see text). *1*, males 20–54 years; *2*, males 17–21 years; *3*, females 19–46 years (derived from data by BURDICK et al. [3])

a given "level" (e.g., the isoelectric line). This goes back to the studies of BURCH et al. [2] and SALTZBERG et al. (1968). Summaries of the methods and examples of their application can be found in ITIL [17] and FINK [10]. Figures 3 and 4 are simplified diagrams demonstrating the basic principles. The EEG signal is examined using a reference line (in this case the isoelectric line), whereby the points at which either the ascending or descending limbs of the EEG signal curve intersect the reference line are considered. The intervals between these points of intersection make up "periods", which are divided into suitable frequency bands of 0.5–3.5 Hz, 3.5–7.5 Hz, 7.5–13 Hz, 13–20 Hz, 20–26.6 Hz, 26.6 Hz–40 Hz, and above 40 Hz. Since the duration of the individual periods is known, one can illustrate in a histogram the percentage that each interval range contributes to the total (Fig. 4).

This method also requires minimal mathematical effort. The result can be easily interpreted if the EEG signal essentially consists of a dominant rhythm. If the signal, however, consists of a complex combination of vari-

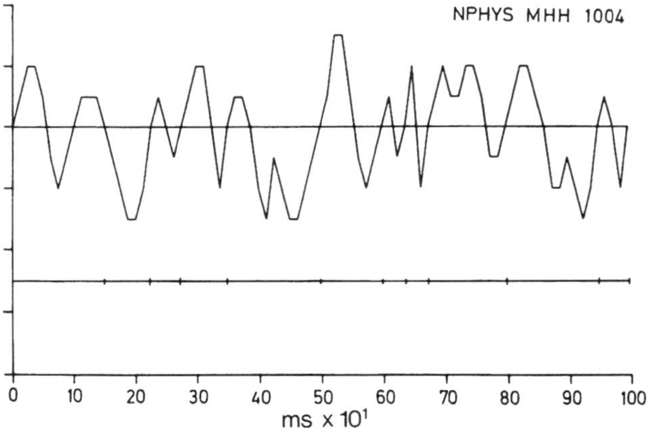

Fig. 3. The principle of period analysis. The *upper part* illustrates the EEG pattern. On the *lower line,* the passages through the level "zero" have been marked; the intervals between them define the periods

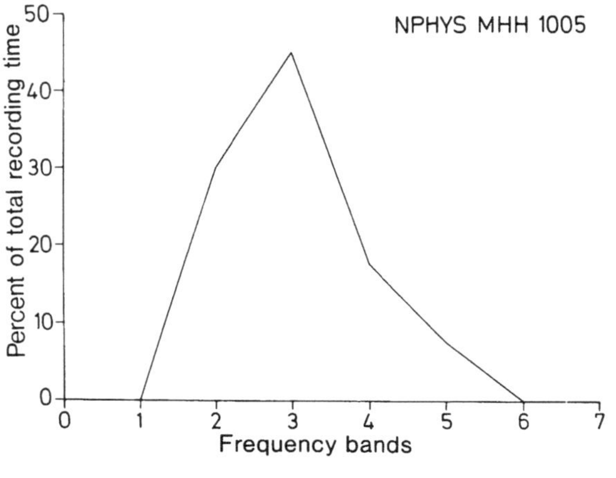

Fig. 4. Frequency profile of the period analysis from Fig. 3. The percentage of the time of registration from the individual groups of intervals or periods is illustrated

able frequencies, incomplete or even erroneous results occur. Figure 3 makes this point clear: Single, rapid wave patterns that are superimposed on slower patterns and do not reach the reference line cannot be detected. Therefore, various modifications of the method have been developed, but they are not completely satisfactory (compare PRONK 1982). It is, therefore, recommended that appropriate frequency ranges are isolated by means of band-pass filtering (according to the frequency bands listed above) and then interval analysis is applied separately to each of these frequency bands [20]. This means that the mathematical effort is multiplied according to the number of bands thus examined and that additional expense is required for exact band-pass filtering. Thus, the former advantage of mathematical simplicity will be lost. A further disadvantage results from the necessarily higher scanning rate of the signal (as required for a digital signal processing), compared with, e.g., spectral analysis of the same signal. It can thus be demonstrated that a scanning rate of approximately 700 Hz (= scanning points per second) is required in order to achieve a frequency resolution of 1 Hz in the frequency band 20–26.6 Hz during interval analysis. During spectral analysis with appropriate low-pass filtering (see below), however, for all clinical purposes, a scanning rate of approximately 80–90 Hz is sufficient, which also has the advantage of less date requiring processing.

Methods of Frequency Analysis

Spectral Analysis. The basics of spectral analysis have been comprehensively and clearly described by DUMERMUTH [5], COOPER et al. (1974), and others, and can be understood without extensive mathematical knowledge.

The following is a discussion of some practical aspects of dealing with EEG spectra; observance of these is most essential for the quality and reliability of the results. The description is limited to the power spectra which are used exclusively here. Further methods, such as cross spectra, coherence functions, and spectra of higher degrees, which also belong to the more general range of spectral analysis, will not be considered. The principle of spectral analysis is based on the fact that every signal, even a nonperiodical one, i.e., here specifically an EEG signal, can be described by a sum of functions with particular properties. Of special practical significance in this context are the trigonometrical functions describing the chronological course of an EEG signal in a formula, as follows:

$$\mathrm{EEG}(t) = a_1 * \sin(w_1 * t) + b_1 * \cos(w_1 * t) + a_2 * \sin(w_2 * t) + b_2 * \cos(w_2 * t) + \ldots$$

EEG (t) signifies the numerical value of the EEG signal at the time t, $w_1 = 2\,\pi/f_1$, $w_2 = 2\,\pi/f_2$, etc. The frequencies f_1, f_2 ... depend on the registration and analytic conditions, e.g., f_1 depends on the duration of the analyzed signal. Practically, a registered EEG signal over several minutes is divided into several segments, e.g., of T seconds. Then it is true that $f_1 = 1/T$. Further, one selects $f_2 = 2*f_1$, $f_3 = 3*f_1$, etc. Already, this clearly shows that the so-called frequency resolution can be determined by an appropriate selection of the length of the segment T. The highest reasonable analyzable frequency, f_N, is limited by the scanning rate $_{\Delta}t$ at which the EEG signal is scanned for digital processing in the computer ("digitized"). For his, $f_N = 1/(2*_{\Delta}t)$ applies. This means that the scanning rate for the signal has to be at least twice as great as the highest frequency under consideration. With spectral analysis of up to 32 Hz, therefore, the scanning rate has to be at least 64 Hz. This simple relation, however, is valid only if frequencies above 32 Hz do not occur in the EEG signal. This cannot simply be assumed in every situation (e.g., in intensive care, line frequency interference of 50 Hz or in muscular artifacts even higher frequencies are encountered). Thus, a much higher scanning frequency would have to be selected than the actual spectral range of interest of 32 Hz in order not to falsify the results of the analysis in an uncontrolled fashion by the higher frequencies in the signal. It is, therefore, more practical and economical to filter the EEG signal prior to digital sampling and processing in such a fashion that the frequencies beyond the range of interest are eliminated (low-pass filtering).

We can now calculate the coefficients a_1, b_1, a_2, b_2 ... This is usually done by means of the so-called fast Fourier transform [4], the algorithmic principles of which need not be discussed here (for details see DUMERMUTH [5]. The so-called spectral coefficients are then calculated as follows:

$$S_1 = \sqrt{(a_1^2 + b_1^2)},\ S_2 = \sqrt{(a_2^2 + b_2^2)}, \ldots$$

Their sequence from f_1 to f_N is the final result of the analysis – the power spectrum. (This illustrates that it would be incorrect to talk of Fourier analysis of the EEG, as the Fourier transform is just an algorithmic step within spectral analysis.) The argument that the brain does not produce sinus waves and that this method is inadequate for EEG analysis is based on a misunderstanding of the nature of spectral analysis. Its application is based on the assumptive axiom – which for continuous basic activity has not been refuted anyway – that it is a stochastic signal occurring randomly in time. Such a signal cannot be exactly determined but can only be described by statistical reference values, e.g., mean or variance. The spectrum merely represents the proportion of this signal as composed by the various frequency ranges; the variability of the signal is resolved into its spectral ranges. (This is the reason for the occasional use of the term "variance spectrum"; since the statistical term variance, in terms of physics, is equivalent to power, the term "power spectrum" is most commonly used.) Under certain conditions – when stationariness and normality exist – this frequency description is sufficient and complete. Stationariness exists when the statistical properties of the signal, especially its variance, remain constant with

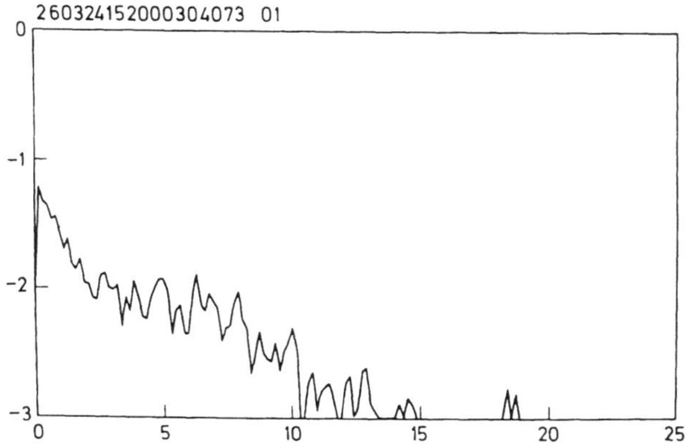

Fig. 5. Power spectrum for a 3-min EEG section, precentral left (*abscissa,* frequency in hertz; *ordinate,* power, logarithmic)

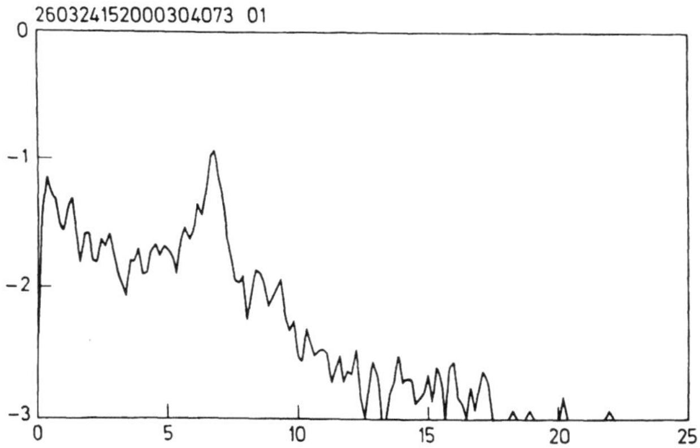

Fig. 6. Power spectrum for a 3-min EEG section, precentral right (*abscissa,* frequency in hertz; *ordinate,* power, logarithmic)

time; normality exists when the amplitude follows a Gaussian (normal) distribution. Both conditions are usually not fulfilled in the case of an EEG, although they are mostly assumed implicitly (compare GASSER [11]). When amplitude distribution deviates from the normal, the spectrum produces an incomplete yet correct description of the statistical properties of the signal. Stationary behavior of the EEG can practically be assumed to last only for a limited period of time (a few seconds); according to individual circumstances and conditions of registration, it can persist for up to 20 s. If a longer segment is chosen for the analysis, an "average" spectrum is obtained for that period of time.

EEG spectra calculated as described do not in general present smooth curves as is demonstrated in Figs. 5 and 6. The spectral coefficients have a relatively low statistical value. This can be improved by smoothing techniques applied to the spectrum. Accordingly, this method leads to a certain decrease in frequency resolution but it is acceptable for the EEG analysis. If in view of the lack of stationariness short segments of, e.g., 4 s are selected, there will be prior to smoothing, according to the above elaboration, a frequency resolution of 0.25 Hz. For practical purposes, this is reduced by the usual smoothing techniques to 0.5 Hz, which can be tolerated without difficulty. Thus, spectra are produced as shown in Fig. 7. A comparable effect is also obtained by averaging a number of spectra from consecutive segments [19]; then, however, the problem of stationariness has to be considered again.

Figure 7 shows a sequence of spectra from consecutive segments arranged in a pseudo-three-dimensional diagram, from left to right corresponding to the temporal left, occipital left, occipital right, and temporal right leads. This kind of "compressed spectral array" [1] allows a synchronous optical display of the EEG dynamics, which would be inaccessible to visual interpretation in such a compact and differentiated form. A conspicuous feature is that the four symmetrically paired channels show a double peak in the alpha range (here at approx. 10 Hz). The observation period covers 30 min; it can be seen how sophisticated the dynamic conditions of basic activity are and how they vary with different channels. Thus, with the aid of spectral analysis, factors can be visualized that cannot be reasonably defined in conventional approaches to clinical electroencephalogy. It also becomes obvious that this form of array is particularly helpful when monitoring EEGs in terms of dynamic aspects – such as in anesthesiological monitoring. It must not be forgotten that an abundant supply of information, which cannot easily be assessed, is offered to the investigator. Therefore, reduction of data is necessary for monitoring and other purposes, i.e., limiting the data to that which is relevant to the problem concerned. At present, this problem is far from being generally solved. One technique that has proven useful for many clinical purposes is based on a parametrization of EEG spectra. It is illustrated in Fig. 8. Here, the division, widely used clinically, of the EEG frequencies into the following frequency bands is employed – delta band (0.5–3.5 Hz), theta band (3.5–7.5 Hz), alpha band (7.5–13 Hz), and beta band (over 13 Hz). Occasionally, it is useful to subdivide the relatively broad beta band into beta-1 (13–18 Hz) and beta-2 (over 18 Hz). The spectral frequency distribution within each of these frequency bands can then be defined by a small number of parameters – first by the integral of the spectrum in each band (which is the area beneath the curve between the band limits, called "absolute band power"). This value, divided by the integral of the whole spectrum (the area beneath the complete curve), can be labeled "relative band power." The maximum of frequency distribution in each band represents the mode, also called the dominant frequency. A measure of the frequency variability within a band is represented by the difference between the 10th and 90th percentiles of the

Fig. 7. Dynamics of the EEG activity over 30 min. *Left* to *right:* temporobasal left, occipital left, occipital right, temporobasal right

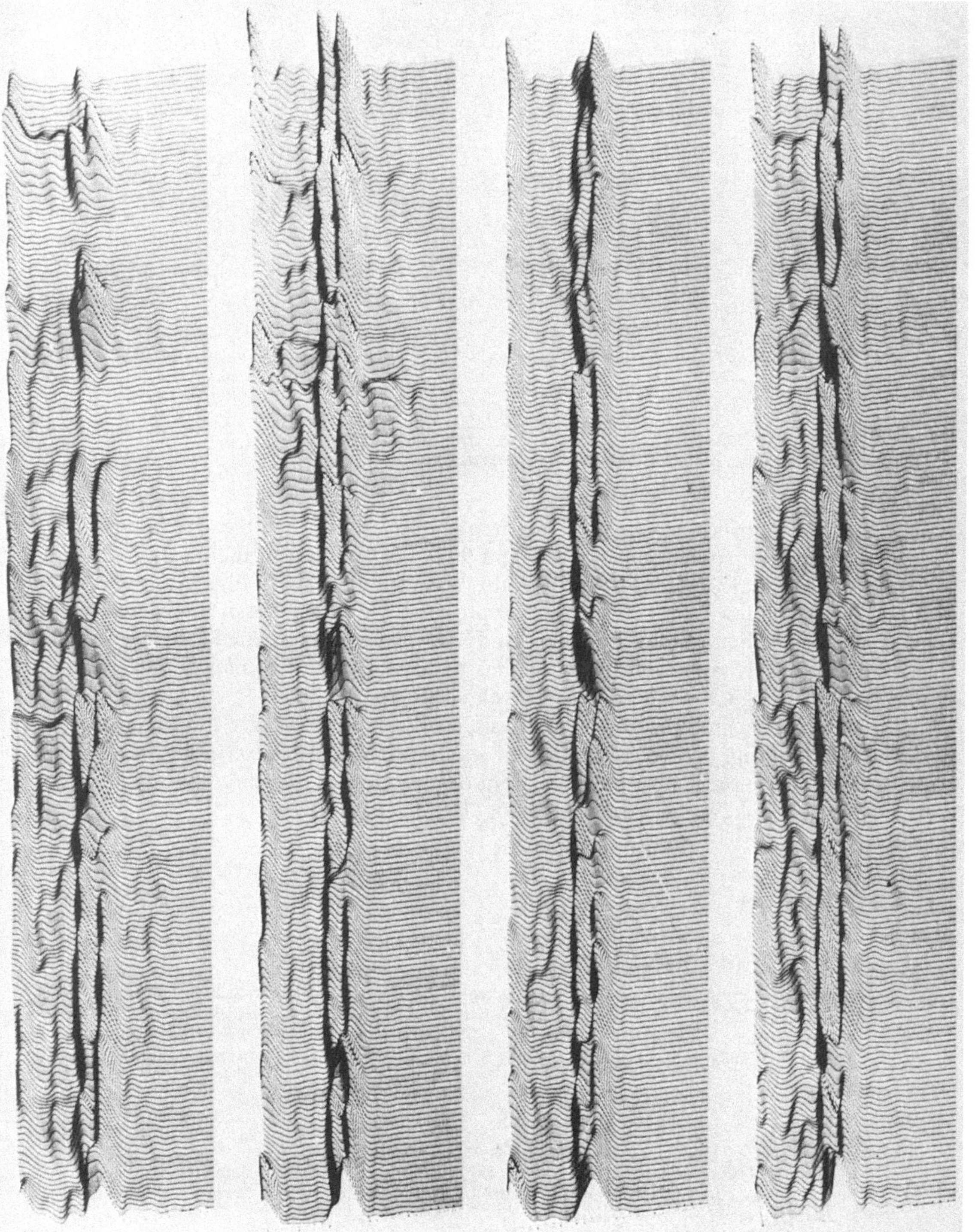

090248171001900575 04 AU 9 090248171001900575 04 AU 10 090248171001900575 04 AU 11 090248171001900575 04 AU 12

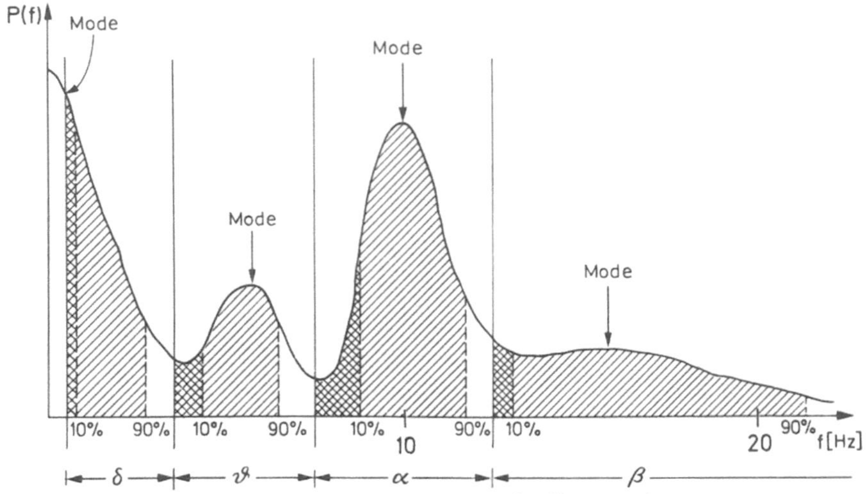

Fig. 8. Definition of the EEG spectral parameters (details see text)

spectral distribution (these are the frequencies up to which the area beneath the spectral pattern reaches 10% and 90%, respectively, of the total area of the particular frequency band). In this way, the diagram in Fig. 7 can be reduced to a few lines representing approximately the course of the power in the delta, theta, alpha, or beta band; however, differentiated information about the dynamics of the course has to be sacrificed. For monitoring purposes, though, the method has proven useful.

Other techniques of EEG analysis, such as autoregressive filtering, Kalman filtering, or the calculation of standardized slope descriptors [15, 16] have been used only rarely for monitoring purposes. Further discussion of these aspects can be found in PRONK (1982).

References

A. Textbooks and Summaries

Cooper R, Osselton JW, Shaw JC (1974) Electroencephalography. Fischer, Stuttgart

Pronk RAF (1982) EEG Processing in cardiac surgery. Institute of Medical Physics TNO. Utrecht Report R-1982-1

Saltzberg B, Edwards RJ, Heath RG (1968) Synoptic analysis of EEG signals. Pergamon, Oxford

B. Papers

1. Bickford RG, Fleming N, Billinger T (1971) Compression of EEG data. In: Trufant SA (ed) Transactions of the American Neurological Association, vol 96. Springer, Berlin Heidelberg New York, p 118–122
2. Burch NR, Nettleton WI, Sweekly J, Edwards RJ (1964) Period analysis of the electroencephalogram on a general-purpose digital computer. Ann NY Acad Sci 115:827–842

3. Burdick JA, Sugermann AA, Goldstein L (1967) The application of regression analysis to quantitative EEG-analysis in man. Psychophysiology 3:249–254
4. Cooley JW, Tukey JW (1965) An algorithm for the machine calculation of complex fourier-series. Math Comp 19:267–301
5. Dumermuth (1973) Numerical spectral analysis of the electroencephalogram. In: Matousek M (ed) Handbook electroencephalography and clinical neurophysiology, vol 5, part A. Elsevier, Amsterdam, p 33–60
6. Drohocki Z (1948) L'intégrateur de l'électroproduction cérébrale pour l'électroencéphalographie quantitative. Rev Neurol (Paris) 80:619–624
7. Etevenon P (1977) Etude méthodologique d'électroencephalographie quantitative. Application à quelques examples. Thèse, Paris
8. Ferber G (1982) Synaptic pattern recognition in clinical routine-EEG. Proceedings of the 6th international conference of pattern recognition, Munich 1982. Computer Society Press, Silver Spring, p 1186
9. Ferber G, Künkel H (1979) Die Bedeutung der automatischen EEG-Analyse für die klinische EEG-Befundung. Wiss Z Ernst-Moritz-Arndt-Univ Greifs. Math Naturwiss Reihe 28:25–36
10. Fink M (1977) Quantitative EEG analysis and psychopharmacology. In: Remond A (ed) EEG informatics: A didactic review of methods and applications of EEG data processing. Elsevier, Amsterdam, p 301–318
11. Gasser T (1977) General characteristics of the EEG as a signal. In: Remond A (ed) EEG informatics. A didactic review of methods and applications of EEG data processing. Elsevier, Amsterdam, p 37–55
12. Goldstein L (1975) Time domain analysis of the EEG: The integrative method: In: Dolce G. Künkel H (eds) CEAN-computerized EEG analysis. Fischer, Stuttgart, p 251–270
13. Gotman J, Gloor P (1976) Automatic recognition and quantification of interictal epileptic activity in the human scalp EEG. Electroencephalogr Clin Neurophysiol 41:513–529
14. Harner RN (1977) EEG-analysis in the time domain. In: Remond A (ed) EEG informatics: A didactic review of methods and applications of EEG data processing. Elsevier. Amsterdam, p 57–82
15. Hjorth B (1970) EEG analysis based on time domain properties. Electroencephalogr Clin Neurophysiol 29:306–310
16. Hjorth B (1973) The physical significance of the time domain descriptors in EEG analysis. Electroencephalogr Clin Neurophysiol 34:321–325
17. Itil TM (1975) Digital computer period analysed EEG in psychiatry and psychopharmacology. In: Dolce G, Künkel H (eds) CEAN-computerized EEG analysis. Fischer, Stuttgart, p 289–308
18. Jacob H (1976) Ein Beitrag zur automatischen in Echtheit ablaufenden Analyse von kontinuierlichen und intermittierenden Aktivitäten im Elektroencephalogramm. Dissertation, TU Hannover
19. Künkel H (1972) Die Spektraldarstellung des EEG. EEG-EMG 3:15–24
20. Künkel H, Westphal M (1970) Quantitative EEG-analysis of Pyrithioxine Action. Pharmakopsychiatr Neuropsychopharmakol 3:41–49
21. Lopes da Silva FH, van Hulten K, Lommen JG, Storm van Leeuwen W, van Vellen CWM, Vliegenthart W (1977) Automatic detection and localization of epileptic foci. Electroencephalogr Clin Neurophysiol 43:1–13
22. Matousek M, Ayvidson A, Friberg S (1979) Serial quantitative electroencephalography. Electroencephalogr Clin Neurophysiol 47:614–622
23. McGillivray B (1977) The application of automated EEG analysis to the diagnosis of epilepsy. In: Remond A (ed) EEG informatics. A didactic review of methods and applications of EEG data processing. Elsevier Amsterdam, p 243–261

III. Technical Requirements for Electroencephalography in the Operating Area

CONTENTS

Under standard conditions, the electric brain activity recorded from the scalp has a potential difference of about 10–100 μV, occasionally up to 300 μV, with a frequency of approximately 0.5–40 Hz; it thus encompasses a considerable dynamic range. This fact and the necessity to suppress any interfering voltage with a possibly high amplitude demand high standards of quality and sensitivity of EEG amplifiers. Therefore, in clinical neurophysiology, EEG recordings are performed in specially fitted and shielded rooms. In addition to comfortably fitted chairs for a sitting or a reclining position – so as to diminish biological artefacts during the tracing – these rooms have special electric equipment, mainly a Faraday cage, to eliminate interfering electric fields. To profit from EEG recordings in anesthesiology, the highly sensitive equipment has to be moved into the operating rooms, where there are a great number of sources of interference that have to be tolerated. Of course, the electric shielding usually used in the EEG laboratory is absent in the operating room. Redesigning the operating rooms is usually not possible for financial reasons. Since the advantage of continuous EEG recording during anesthesia is of immense value, this chapter discusses means of obtaining EEG recordings despite the above-mentioned obstacles ([3] FAULCONER and BICKFORD 1960; BRECHNER and WALTER 1962; SADOVE et al. 1967).

Electrodes

The tracing of cerebral electric activity from the scalp, excluding invasive methods, may present a problem. Figure 1 shows several kinds of electrode in use. In the neurophysiological field, with alert and cooperative patients, *pad or z electrodes* are used, fixed to the convexity of the skull by a rubber cap (Figs. 1, 2). Their application in anesthesiology cannot be recommended because they are detached easily during manipulation of the head.

Needle electrodes are pricked into the scalp. Their advantages are quick application and high stability of registration quality. Considerable disadvantages are skin lesions and painful application. Needle electrodes should, therefore, be used only during anesthesia. Since, however, an initial EEG prior to anesthesia allows a better evaluation of intraoperative changes, needle electrodes are rarely applied.

Adhesive electrodes are regarded as optimal during surgery. They consist of a cuplike silver/silver chloride electrode firmly soldered on to a wire and they have a plastic rim for fixation. These electrodes are attached to the scalp with collodion; with a little practice it is possible to attach them without removing any hair. After fixation, the cup of the electrode is filled with electrode cream. The advantages of adhesive electrodes are mechanical and electric stability, low transition resistance, and noninvasive application. However, their application requires time and skill. In neurophysiology, such electrodes are used for recordings during sleep (Fig. 3). Other kinds of electrode developed for special neurological studies are of no significant value for anesthesiology; they will not, therefore, be discussed further.

The *electric properties* of the various *electrode materials* affect the quality of the recorded track. In practice, the quality of the combination silver/silver chloride has proven most favorable for EEG recording. The combination of the precious metal and its salt provides the electrode with a low

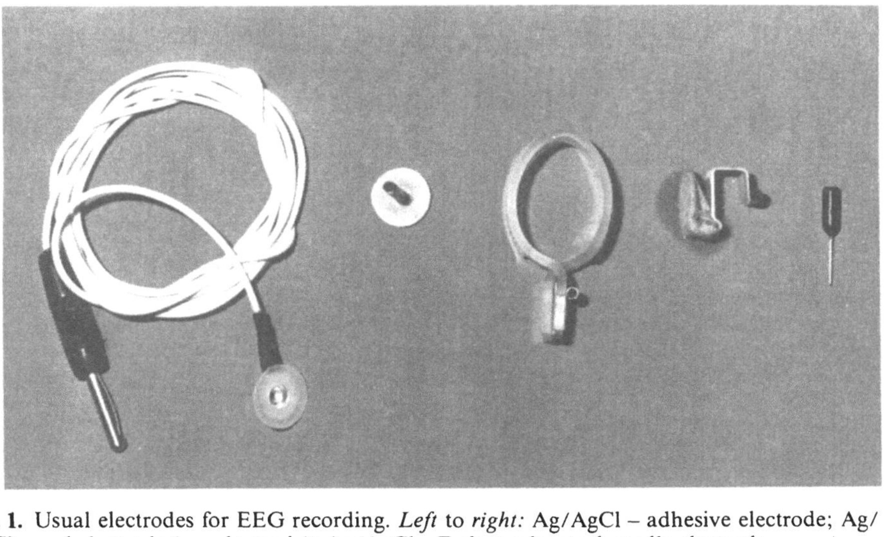

Fig. 1. Usual electrodes for EEG recording. *Left* to *right:* Ag/AgCl – adhesive electrode; Ag/AgCl – pad electrode (ear electrode); Ag/AgCl – Z electrode; steel needle electrode

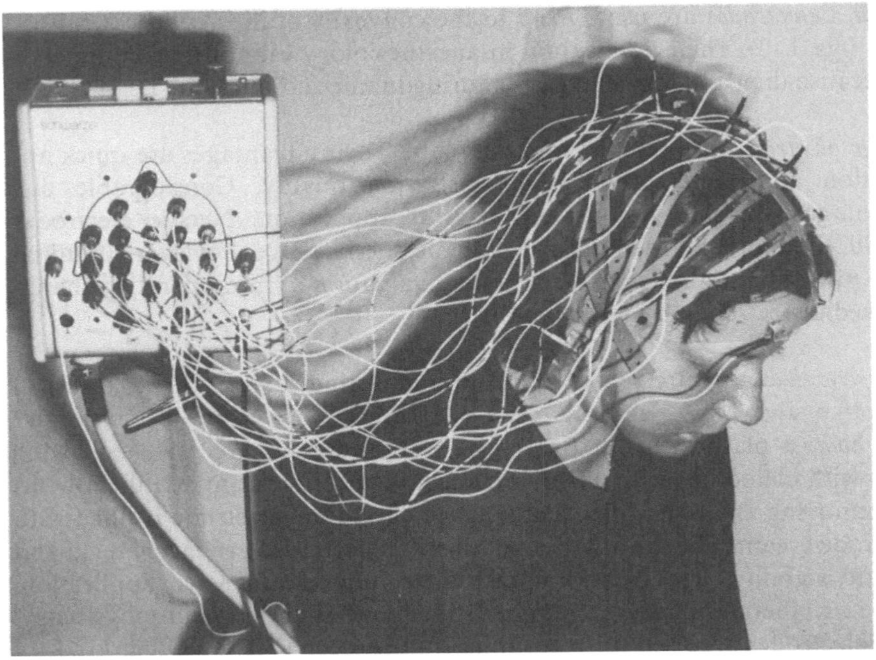

Fig. 2. Pad and Z electrodes are attached to the scalp by means of a rubber cap; the usual method of recording in alert patients in routine clinical practice

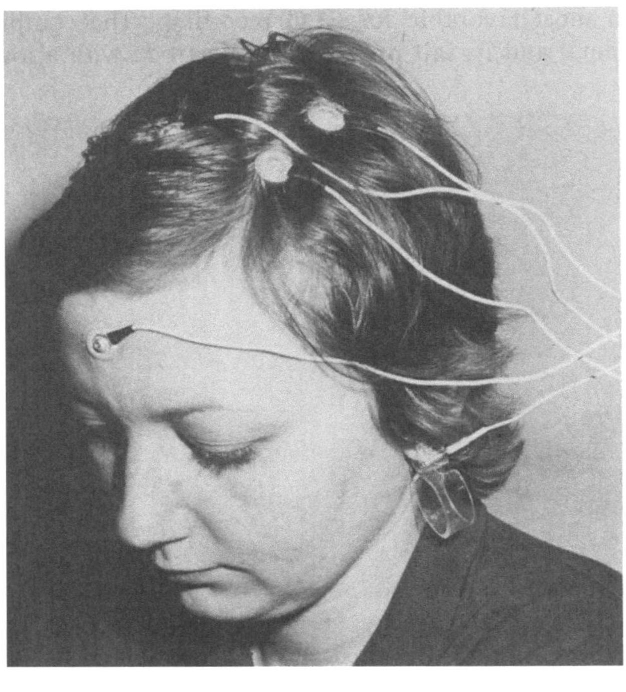

Fig. 3. Method of recording in our patients during surgical anesthesia – adhesive scalp electrodes for two tracks; C_Z–A_1 and C_3–P_3 (frontal electrode = earth electrode)

transition resistance, even for low-frequency interference, because the condensatorlike properties that occur with plain-metal electrodes are here compensated when the electrodes are attached to the head of the patient. The mechanical stability as well as the transition resistance have to be observed. For this purpose, most EEG machines have a special device for measuring the so-called electrode impedance. The impedance should be approximately 10 kΩ; it should definitely be below 50 kΩ and should be distributed as equally as possible when using several electrodes, otherwise considerable interference could occur. Careful application of the electrode is perhaps somewhat time-consuming, but is essential for a clear recording.

Amplifier and Recorder

The choice of the EEG machine mainly depends on personal taste and financial resources. The EEGs in this book were recorded on machines made by the Schwarzer company. Three parameters have to be considered with regard to an EEG machine. These parameters can be modified for all channels simultaneously and individually for each channel, which results in a large number of operating switches on the apparatus. These three parameters will now be discussed:

1. *Amplification.* Generally, one begins with an amplification of 50 μV = 7 mm. This means that an input signal of 50 μV will be amplified until the deflection of the recording needle on the paper is 7 mm. Conventionally, EEG machines are constructed so that a positive input voltage produces a downward deflection of the needle. Because anesthesia may lead to a very flat recording, optionally higher amplifications, e.g., 10 μV = 7 mm, are desirable.

2. *High-frequency Filters.* These filters limit the upper frequency of the recording. The main frequency range of electric brain activity is 0.5–40 Hz. Usually, tracings are performed with the filter set at 70 Hz. It may be necessary to lower the upper limit to 30 Hz, especially in the operating area. This may lead to a loss of information and this change must be noted on the recording.

3. *Time Constant.* The time constant is a parameter given in seconds of determining lower frequency limit. The term "time constant" results from the working mode of such filters: If a quick-changing function is fed into the filter, i.e., the voltage "0" is abruptly brought to a constant d c level, there will initially be the same jump on the output of the filter. Subsequently, the voltage decreases exponentially to "0". The time constant is the period during which the voltage is reduced to 37% of its initial level. The values of lower frequency limit (F) and time constant (T) are related as follows: $F = 1\ (2\ \pi\ T)$. Typical levels that can be set are – 0.003 s, 0.1 s, 0.3 s, 1 s, or

expressed in terms of frequency – 5.3 Hz, 1.6 Hz, 0.53 Hz, and 0.16 Hz. A conventional EEG is usually recorded with a T of 0.3 s. By decreasing the T, slow interference due to perspiration or respirator activity can be suppressed. Note, however, that an EEG phenomenon of the delta range (0.5–3.5 Hz), which is essential for anesthesia, may be partly suppressed by increasing the lower frequency limit. These three important parameters should be adjusted in a like manner for all channels. On most machines, there are additional switches for adjusting all tracks simultaneously.

In Germany the speed of paper transport is usually 30 mm/s; in special cases different speeds may be selected (a summary of the technical and physical requirements for EEG recordings can be found in COOPER et al. 1974). The EEG machine is attached to a connection box, into which the banana plugs of the electrode cables are inserted. This cable is usually 2.5–3 m long. Nowadays, recorders with a miniaturized input amplifier integrated into the connection box are widely used. The EEG signal is unamplified only along the electrode cables; this limits the influence of interfering electric or magnetic fields to a minimum. Using such devices, the distance between the connection box and the EEG machine can be increased to a certain extent. Owing to the lack of space in the operating room, the recorder can be placed outside, however, a disadvantage is that the anesthesiologist cannot then observe the actual tracing. An additional monitor, as used in recording EEG, could be employed. More recently, so-called electrode amplifiers have been developed, with electrodes and amplifier integrated by means of microminiaturization. Thus, the signal can be amplified at the patient's head, resulting in improved protection against interfering factors.

Processing of EEG Signals by Spectral Analysis

Electroencephalographic recordings in the conventional manner, described above, without further analysis, could be sufficient for use in anesthesiology. The EEG curve, however, is such a complex pattern of different frequencies, wave forms and voltages that a physician with a good deal of EEG experience would be required for the assessment. A great aid to the anesthesiologist in interpreting the EEG, therefore, is computerized EEG analysis by the on-line technique. There may be a loss of information compared with the original recording, mainly in the recognition of certain EEG patterns or pathological waves. Since in anesthesiology this is only of importance in special situations, the loss of information can generally be tolerated during anesthesia.

Of the large number of options for computerized EEG interpretation, spectral analysis supported by the fast Fourier transform has proven to be suitable in practice and has also been successfully applied by us. Details of this method are described elsewhere (Part A, Chap. II). Spectral analysis transforms the EEG signal recorded on the time axis into a signal on the frequency axis. The resulting power spectrum demonstrates the electric power of the various frequency ranges for the time analyzed. These spectra are suitable both for visual interpretation and numerical comparison. If several spectra of consecutive EEG signals are recorded one above another, a "spectral mountain" is obtained, allowing a very clear demonstration of any changes in the alpha rhythm, e.g., as a result of the effects of drugs (Figs. 4, 5). The spectral analyses presented in this book were recorded using the biosignal processor BIO 16 made by AEG Telefunken with the so-called LEM (long-term EEG analysis) program. This analysis involved in each case a 16-s portion of the EEG curve; 14 s were then required for printing the results. This entailed that the spectra were recorded at 30-s intervals. All the figures in this book were drawn using the same program. The graphs were produced by an X/Y plotter. The machine can analyze two EEG tracks simultaneously.

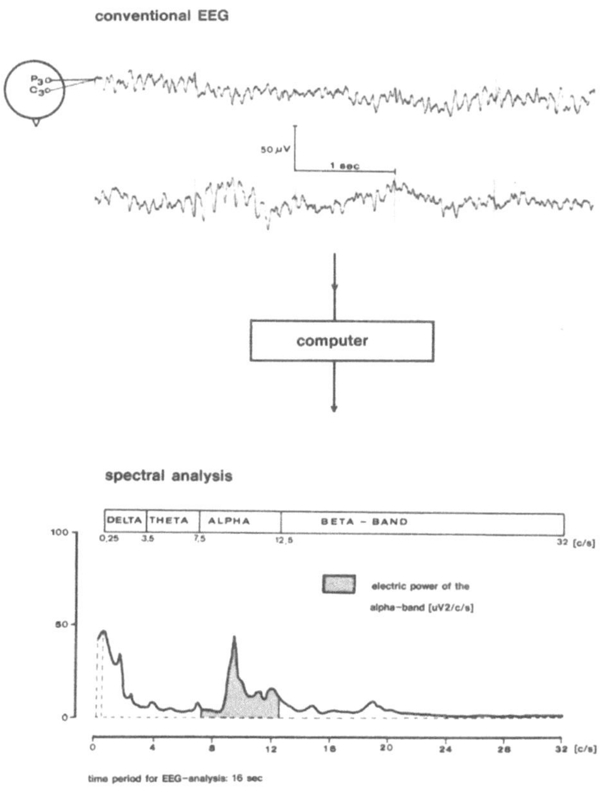

Fig. 4. Schema of the transfer of the EEG from time to frequency range. *Top* joint EEG lead, the electric tension is recorded in relationship to the time. *Bottom* through the fast fourier transform (FFT) a "spectral mountain" of the electric power in relationship to the frequency is formed in the computer

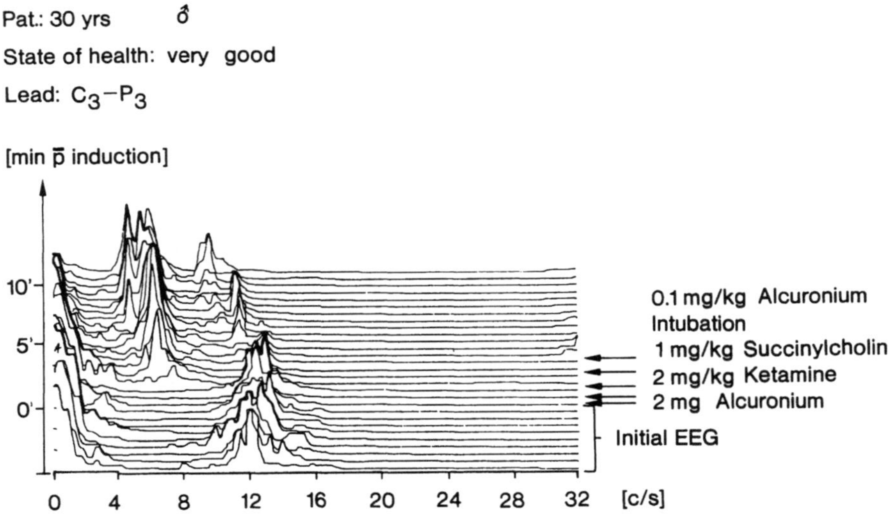

Fig. 5. Sample case of a 15-min EEG spectral analysis prior to and after i.v. application of ketamine for induction of anesthesia. The changes in the dominant frequency can easily be observed in chronological order as "plastic spectral mountains"

Safety Precautions

The recording of an EEG is an absolutely harmless method for the patient and is devoid of any side effects if the electric appliances are intact and have been serviced properly. The instructions of the operating manual must be followed. When employing the set in the operating room, the extensive regulations of the publications "Safety of Electric Systems"

IEEE Std 134–1965
IEEE Std 128–1976

"Maintenance, Testing and Radiation"

IEEE Std 141–1976
ANSI N 449–1–1978

have to be followed.

Compliance with these regulations will render continual EEG recording during anesthesia safe for the patient. When high-frequency currents are used by the surgeon, the same precautions as with continual EEG recording apply.

Artefacts: Recognition and Avoidance

Due to the low voltage of an EEG signal, recognizing and avoiding interfering artefacts is essential to the [2] diagnostic value of the recording. Severe interference (e.g., high-frequency coagulator) is easily recognized –

Table 1. Biological artefacts in EEG

Cause of artefact	Avoided by	Remarks
Muscular movements (blinking, eye movements, muscular tremor etc.)	Talking with the patient; patient adopting comfortable sitting or reclining position; asking patient to lie calmly	Artefacts due to movements usually disappear after induction of anesthesia and muscle relaxation. More pronounced movements may simulate activities in the delta-band
Cardial artefacts (R peak in EEG)	Changing position of reference electrode	Cardial artefacts are of little relevance if recognized early. In spectral analysis a peak of 1–2 Hz may be seen
Pulse-wave artefacts	Changing position of electrode if electrode was placed over artery	–
Perspiration	Adequately ventilating recording room. If anesthesia is too shallow it should be deepened	Respiratory artefacts are very slow potentials, suppressible by changing the time constant, leading to a loss of information in the delta band
Electrodes and cables	Decreasing electrode resistance (less than 50 kΩ). Keeping head still – even during anesthesia (head should be limited)	Transition of biological to technical artefacts

Table 2. Technical artefacts in EEG

Cause of artefacts	Avoided by	Remarks
Interference from 50-Hz ac (electric mains) "electric buzzing"	1. Excluding faulty equipment 2. Reducing impedance of electrodes 3. Modifying the room if interference is due to electric or magnetic fields 4. Separating cables for recording and power supply 5. Compensating different potentials, when several electric sets are used on one patient – "central earthing"	In modern equipment with high electrode impedance, buzzing can almost always be controlled by a 50-Hz filter
High-frequency artefacts e.g., due to cautery, neon lamps	Removing the cauterant usually impractical. Replacing tube or transformator. Screening all electric lights	Caution: during cauterization, EEG recorder must be in the "bloc" position, otherwise recording pens may get damaged
Cable artefacts through general disturbance in the recording room	Reducing number of people in recording area; relaying cables in protected areas	Reducing the number of people is not always practical

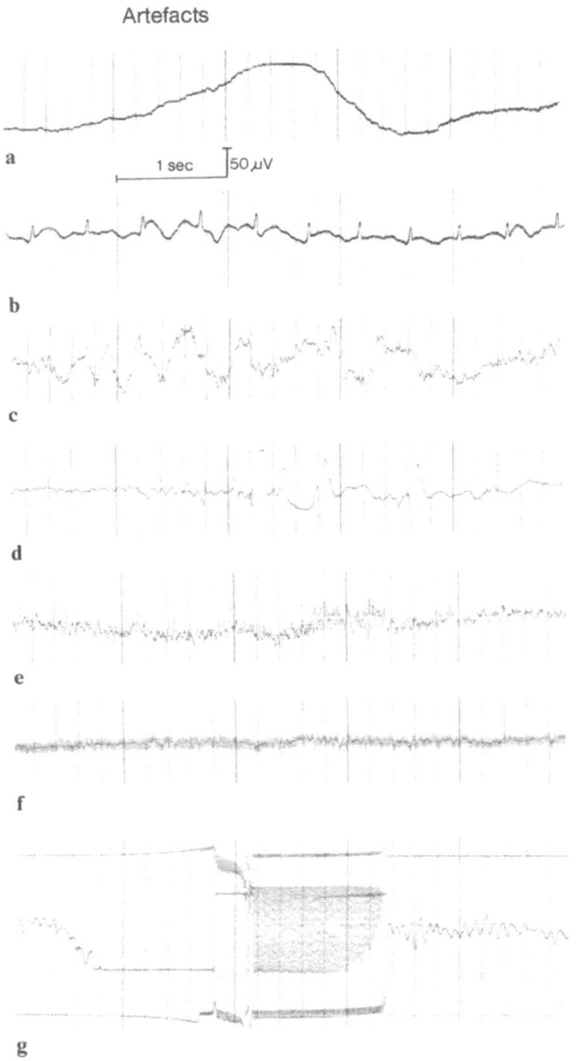

Fig. 6 a–g. Frequent interferences in the conventional EEG. **a** sweating; **b** R peaks of the ECG; **c** movement by the patients (easily confused with the delta activity); **d** cable artefacts; **e** muscular tremor (occasionally simulating fast beta activity); **f** 50 Hz interference ("power buzzing"); **g** interfering surgery (coagulator)

even by the less experienced. There are, however, artefacts that simulate abnormal cerebral electric activity and have to be recognized. This is essential in avoiding errors, both diagnostic and therapeutic. Artefacts are classified into two groups – biological (Table 1) and technical (Table 2). Biological artefacts are produced by the patient himself or occur between the patient and the electrode; technical artefacts are caused by the recorder or other equipment. Tables 1 and 2 and Fig. 6 list the most frequent artefacts and measures to avoid them.

Selecting the Recording Lead System

The simultaneous recording of several EEG tracks has resulted in the development of a large number of different recording patterns, the advantages and disadvantages of which will be discussed below. Thus, each EEG laboratory tends to prefer its own recording pattern (CHRISTIAN 1975, KUGLER 1981). A certain standardization – at least of the site of the leads –

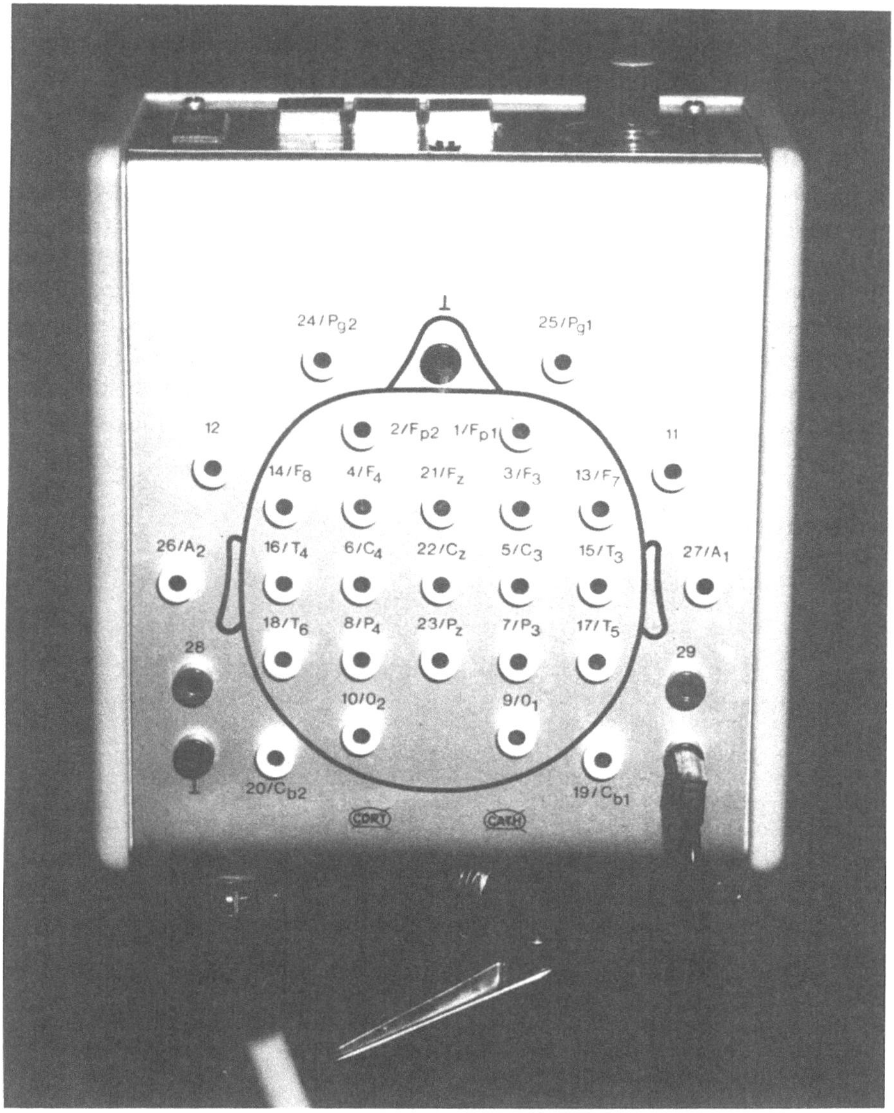

Fig. 7. Pickup head of the EEG recorder. Electrode positions are labeled according to the international 10/20 system

was achieved with the introduction of the international 10/20 system (Fig. 7). The following guidelines should be observed in order that recordings from different study-groups may be compared:

1. *The placement of the electrodes should be in accordance with the international 10/20 system.*
2. A complete EEG in the neurophysiological sense is technically time-consuming. Thus, *in routine* surgery and *anesthesia, the number of tracks must be kept to a minimum.* When selecting the tracks, the positioning of the electrodes should be such that it gives rise to little mechanical strain. Topographical peculiarities of the normal background activities have to be taken into account (see also Part A, Chap. IV).
3. *The recording pattern, once chosen, should be maintained.*

In this book, all the recordings were made with two tracks – left precentral-parietal (C3-P3) and vertex left earlobe (CZ-A1). The reduction to two tracks was initially made for technical reasons, as only two tracks could be analyzed simultaneously. This limitation has proven to be tolerable since in anesthesiological routine work mostly neurologically healthy patients are studied by means of EEG. In the case of focal EEG changes in the "alert EEG," one has to determine individually to what extent EEG monitoring should be carried out with regard to possible diagnostic or therapeutic measures.

References

A. Textbooks and Summaries

Brechner V, Walter R (1962) Practical Electroencephalography for Anesthesiologist. Thomas, Springfield

Christian W (1975) Klinische Elektroenzephalographie. Thieme, Stuttgart

Cooper R, Osselton JW, Shaw JC (1974) Elektroenzephalographie. Fischer, Stuttgart

Faulconer A, Bickford R (1960) Electroencephalography in Anesthesiology. Thomas, Springfield

Kugler J (1981) Elektroenzephalographie in Klinik und Praxis, 3rd edn. Thieme, Stuttgart

Sadove MS, Becka D, Gibbs FA (1967) Electroencephalography for Anesthesiologists and Surgeons. Pitman, London

B. Papers

1. Streu BK (1977) Elektrische Sicherheit im Krankenhaus. Hellige Beiträge für die Medizin, vol. 4. Hellige, Freiburg
2. Weber PA (1976) Mögliche elektrische Störungen bei der Gewinnung des EEG. J Electrophysiol Technol 2: 172–186
3. Wiemers K, Puppel H (1959) Praktische Bedeutung der EEG-Registrierung im Operationssaal. In: Weber A, Heilmeyer L (eds) (Hrsg) 2. Freiburger Kolloquium über Kreislaufverhalten. Werk, München, pp 25–54

IV. Initial Electroencephalographic Findings in Patients Undergoing Anesthesia

CONTENTS

Findings in the Average Population

The EEG basic or background activity is defined as the continuous sequence of waves that predominate in frequency and amplitude (8–16 leads). Classifying the usually recorded frequency range of 0.5–32 Hz into four frequency bands (Table 1) has proven useful in the visual evaluation of the conventional EEG. However, the lack of a sharp transition between the defined ranges frequently makes such an arbitrary division questionable, e.g., when the actual level of consciousness changes through various causes (KÜNKEL 1980).

Between early infancy and the end of puberty, the EEG shows several developmental stages and pronounced EEG changes (DUMERMUTH 1976). In the adult, the individual EEG type can be clearly determined; until late middle age it shows a certain consistency [12]. The individual EEG type has thus far been shown to be only partly hereditary [15, 16, 17, 20]. A continuous change occurs again later in life and is widely variable in its individual extent; it is mainly due to the metabolic and hemodynamic processes of aging [9, 12]. Although in the EEG of a healthy adult parts of all frequency bands are frequently seen, one or two of the "classic frequency ranges" – alpha, beta, or theta – predominate. The individual background activity of the EEG types (Table 2; Figs. 1–5) is determined accordingly (JUNG 1953; CHATRION and LAIRY 1976; KÜNKEL 1980).

Table 1. Frequency bands of the EEG between 0.5 and 32 Hz

Frequency band	Frequency (Hz)
Delta	0.5– 3.5
Theta	3.5– 7.5
Alpha	7.5–12.5
Beta	> 13

Table 2. EEG types of the electroencephalographic basic activity according to JUNG (1953) and KÜNKEL (1980)

EEG type	Dominant frequencies (Hz)	Voltage (μV)
Alpha	8 –13	10–50
Beta	12.5–32	10–50
Partial beta	8 –13 + 12.5–32	10–50 5–30
Irregular	4 –32	< 100
Low-voltage	dominant frequency poorly visible	

Table 3. Basic EEG patterns

EEG type	Dominant frequency ranges in occipital region (Hz)
Alpha	8–13
Beta	16–25
Theta	4– 8
Combined forms	2–32 without a dominant frequency

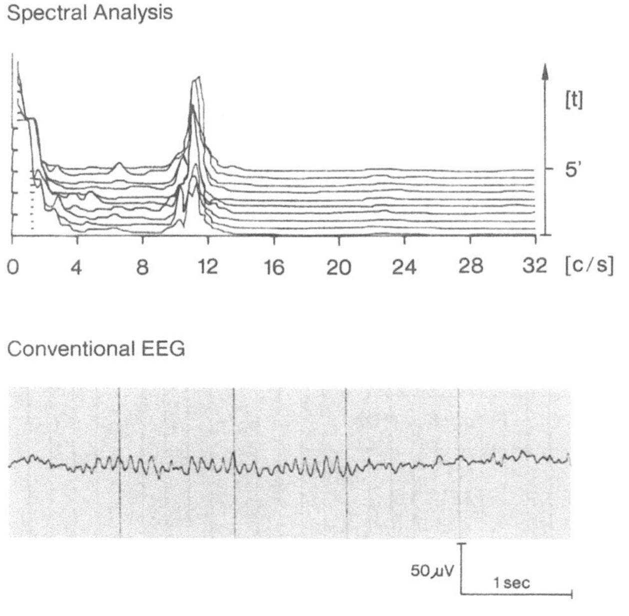

Fig. 1. Alpha EEG in *top* spectral analysis and *bottom* in the conventional manner.
Lead: C_3–P_3

The EEG types of the background activity as listed in Table 2 occur in 95% of the adult population and thus represent the most frequent variants of the EEG (KÜNKEL 1980). A simplified classification of the background activity types was described by KUBICKI and HOELLER [8]. Decisive for the type definition is the dominant frequency over the occipital region (Table 3). Wave patterns in other regions of the brain do not affect the determination of the EEG type.

The classification according to potential difference uses the terms high (> 50 μV), normal (10–50 μV), low (3–10 μV), and extremely low (< 3 μV).

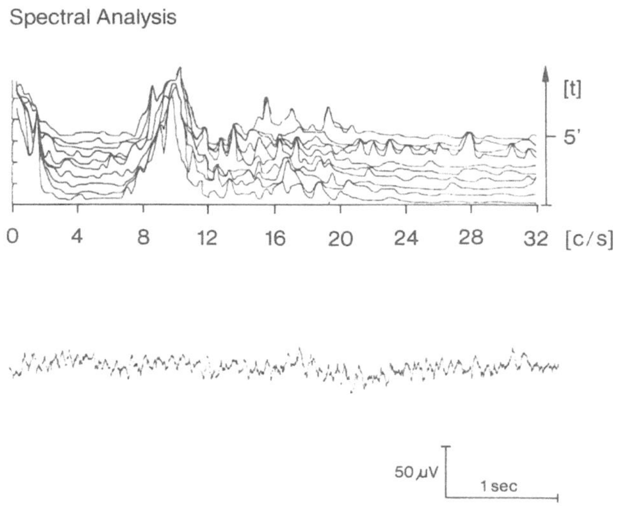

Fig. 2. Alpha + beta EEG. Lead: C_3–P_3

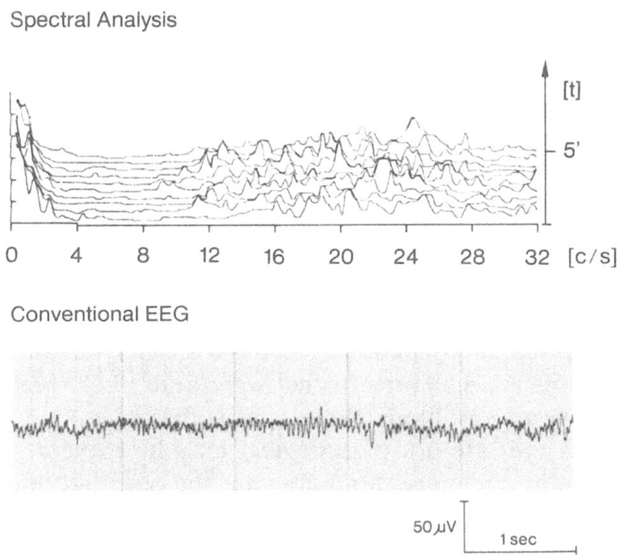

Fig. 3. Beta EEG. Lead: C_3–P_3

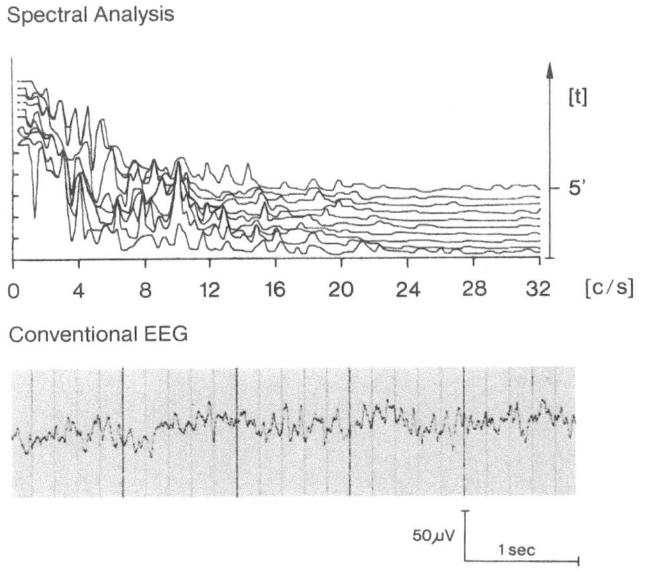

Fig. 4. Irregular EEG. Lead: C_Z–A_1

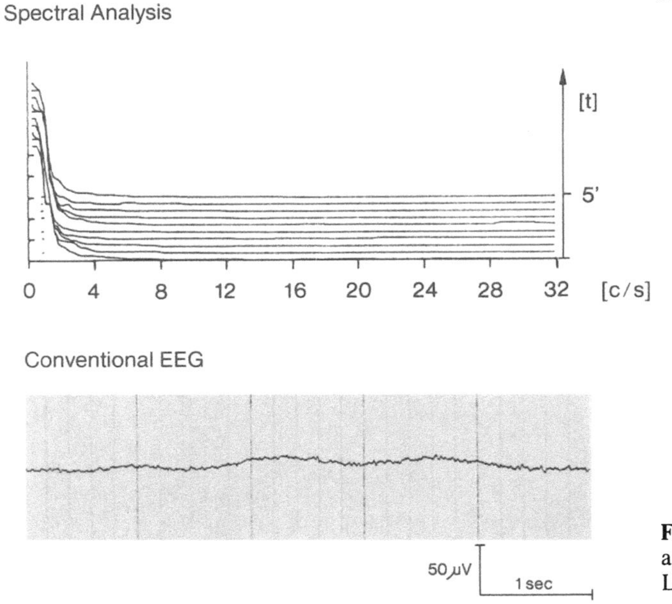

Fig. 5. Flat EEG, low-voltage EEG. Lead: C_3–P_3

Approximately 82% of the basic EEG types fall into the normal range of 10–50 µV, 4.5% are in the high, and 13.8% are the low or extremely low range of amplitude. In an individual case, one, occasionally several, variants that are not pathological may be present in the background activity. Among them are the following: the occipital theta variant, the mu activity, the precentral beta groups, the lambda waves, the so-called sleep-lambda, and the kappa waves.

Whereas the EEG in the normal range has a wide interindividual variability, findings in the same person remain generally constant between the ages of 20 and 50 years. Studies of the distribution of EEG findings in the standard population (Tables 4, 5) show an alpha type in approximately 65–90% of cases; the pure beta type is rare – approximately 7% of cases. Findings cited in the literature concerning the occurrence of beta activity

Table 4. Distribution of basic EEG types in standard population

EEG type	Incidence (%)	Author
Alpha	75–90	Gibbs et al. ([6] 1950, 1971)
	88–90	Vogel et al. [18, 19, 21]
	86	Kubicki and Hoeller [8]
Beta		
(13.5–17.5 Hz)	3.6	Brazier and Finesinger [2]
(17.5 Hz)	100	
	6.6	Gibbs et al. [6]
	3.0	Finley [4]
	47	Mundy-Castle [11]
	82–97	Mundy-Castle et al. [12]
	51	Obrist [13]
	0.0	Gallais et al. [5]
	28	Picard et al. [14]
	8.2–33.3	Roger and Bert [17]
	8.1 } 5.2– 8.1 }	Vogel and Fujiya [20]
Theta and irregular	5 – 7	Gibbs et al. ([6] 1950)
	1 – 0.1	Vogel et al. [18, 19, 21]
	5.6	Kubicki and Hoeller [8]
Low-voltage	9	Gibbs et al. [6]
	4.1	Cohn (1949)
	11.6	Gibbs and Gibbs (1950)
	7.3	Pine and Pine [15]
	19	Gallais et al. [5]
	12	Picard et al. [14]
	8	Adams [1]
	13–14	Kubicki and Hoeller [8]

Table 5. Distribution of EEG types in old age (65–94 years)

Alpha	65–51	Obrist [13]
7–8 Hz variant	17–36	Obrist [13]
Beta	12 ca.	Obrist [13]
Partial beta	ca. 51 } 55	Obrist [13]
	ca. 59 }	Simon (1977)
Irregular	9–17	Obrist [13]
Low-voltage	0	Obrist [13]
	1	Simon (1967)
	13.3	Martelli [10]
Delta/theta focal findings	30–50	Obrist and Busse (1965)

vary considerably (Table 4). This is caused by the selection of different sites for the recordings leads and topographical differences in the distribution of beta activity in the brain [7, 12]. Theta EEG background activity, theta variants, irregular theta forms, or generally irregular EEG variants are reported in 0.1–7% of cases. A low-voltage EEG occurs in 4–14% of cases.

Four kinds of EEG change in particular occur with age, causing a considerable shift in the normal distribution (Table 5; for survey of reference material see OBRIST and BUSSE 1965).

1. *The Dominant Alpha Frequency Becomes Slower.* The dominant peak of the alpha frequency between 20 and 50 years of age is 10.2–10.5 Hz. It decreases to approximately 8 Hz at the age of 95 years. Generally, males are more affected by this shift then females. As a pathophysiological correlate of the alpha deceleration, cerebrovascular functional and organic senile changes are assumed to take place. According to the individual degree of displacement of the alpha peak, these changes may not be serious or they may be correlated with the slowing of intellectual skills and memory. This so-called slow alpha-frequency, or the 7–8-Hz variant in the EEG, may be regarded as a first sign of cerebrovascular change, even without any clinical manifestations. The incidence of the 7–8-Hz rhythm in the EEG normal distribution is approximately 17% below the age of 70 years; above 70 years of age it is approximately 36%; the 8–13-Hz EEG accordingly decreases from 86% to 51–65%. At the same time in older people, the overall voltage of the EEG frequencies is lower; the EEG reactivity is slower [3].

2. *Beta Activity Becomes More Frequent.* In 50–60% of older people, mainly females, a partial beta EEG, in 12% of cases a pure beta EEG, is observed. Increases in the beta activity occur after 50 years of age. A further increase in frequency until the 70th year is followed by a decrease. The amplitude usually ranges between 10 and 25 μV. Beta EEG forms in late middle age are correlated with well-preserved intellectual skills. They possibly point to the existence of cerebral compensatory mechanisms, which conteract age-induced decelerations in indeterminable areas of the brain.

3. *Diffuse Deceleration of Frequency Occasionally Occurs.* Irregular EEGs with frequencies of 4–7 Hz and/or 1–3 Hz are often observed – approximately in older age. More or less pronounced stages of decreased intellectual skills and cerebral confusion are associated with such EEG findings. Arteriosclerotic vascular changes, cardiac and circulatory failure, and imbalances of the cerebral blood supply, and thus of the cerebral metabolism are regarded as the causes. A diffuse EEG deceleration is – when pronounced – regarded as unfavorable both for intellectual skills and general life expectancy (KUGLER 1981).

4. *Focal EEG Changes Become More Frequent.* Between 50 and 70 years of age, focal delta or theta waves occur in 30–50% of the population, predominantly left-temporally. No increase in the number of focal findings is seen in later life. Such changes suggest a regionally limited impairment of cerebral perfusion and metabolism. Clinically, they usually remain silent. With further extension of the focal EEG changes, clinical/neurological correlates can be observed.

Findings in Patients Undergoing Anesthesia

EEG Findings. The distribution of the basic EEG activity was derived from 1000 patients (20–50 years of age, n = 750; over 70 years, n = 250) of a general department of surgery and anesthesiology, approximately prior to operation. The subjects studied were between 20 and 90 years of age. Patients with a history of cerebral pathology were excluded. The EEG recordings were performed both on patients who had previously received medical treatment and on those with no history of such treatment. Alertness and initial psychic state were assessed subjectively.

Since the EEG trackings were not recorded from the occipital region, a strict classification of the EEG findings according to the proposals for a systematization of the basic EEG types and variants made by Kubicki and Hoeller [8] was not possible. Therefore, the EEG type definition of Jung (1953) and, in geriatric patients, the definition of Obrist [13] were used.

A comparison of the findings in our own patients aged between 20 and 50 years with those of the normal population (Table 4, 6), shows the proportional distribution of the basic EEG types – alpha, beta, irregular (theta), low voltage – to be similar. Only the considerably high percentage of additional beta activity, which in our patients was 33%, as opposed to a mean of 12%, deviates markedly from the average population. Two factors bring this about: (a) the use of psychotropic substances, which increase cerebral beta activity the night prior to surgery or immediately before surgery (30–45 min); (b) the choice of location of the recording leads (C_3–P_3) in areas of high cerebral beta activity.

The EEG findings in geriatric patients determined in this study (Table 7) are similar to the results of Obrist [13], who reported the proportion of 7–8 Hz variants and the irregular frequency distribution in old people (see Table 5). The considerable differences between the results in the alpha range may be explained by the different forms of group classification. Obrist evidently first listed all individuals showing activity in the alpha range (51–65%). The whole group was then further subdivided into subjects with a 7–8 Hz variant (17–36%) and those with additional beta activity (alpha, beta, or partial beta; approximately 51%). In our study, individuals were only included in the "alpha group" if they presented a pure alpha ac-

Table 6. Distribution of basic EEG types in young patients [$n=750$, age 20–50 years; classification according to JUNG (1953)]

EEG type	Distribution
Alpha/alpha + beta	43/33 (=76)
Beta	3
Irregular	8
Low-voltage	13

Table 7. Distribution of basic EEG types and variants in geriatric patients ($n=250$, age > 70 years; classification according to OBRIST [13])

EEG type	Distribution (%)
Alpha	27
7–8-Hz variant	18
Beta	1
Partial beta	19
Irregular	22
Low-voltage	7
Delta/theta	6

Table 8. EEG findings in geriatric patients ($n=250$, age > 70 years)

1. Normal initial EEG (alpha, beta, alpha + beta-variants)	47
2. Low-voltage initial EEG (variant, fear due to tension)	7
3. Altered but not definitely pathological initial EEG (transition to irregular findings, without determination of frequency ranges, borderline decrease in alpha frequency)	18
4. Initial EEG with pathological changes (dominant frequency 7 Hz, irregular frequency; delta/theta, alpha and delta; delta/theta/alpha/beta-type)	28
Summary:	
a) Normal initial EEG (groups 1 and 2)	54
b) Altered initial EEG (groups 3 and 4)	46

tivity of 8–13 Hz, with no additional frequencies and no slow variants (Table 7). If these subgroups are included, as in OBRIST's study, the alpha type in our own patients is seen to be just as frequent, at 64% (Table 7). In patients with a pure beta type, the clear difference persists between our finding of 1% and the value of approximately 12% in the literature [12]. Also, the percentage of beta variants (alpha plus beta) in our geriatric patients is lower (19%) than that found by OBRIST [12] or SIMON (1977) (55%). The reason for this might be the lower age limit of the patients: OBRIST set the limit at 65 years; for our patients it was 70 years. Low-voltage EEG patterns, due to psychic tension prior to surgery were more frequent in our patients. Patients over 70 years old undergoing anesthesia show a normal initial EEG in approximately 54% and a more or less altered EEG pattern in approximately 46% of cases (Table 8).

Interpretation of EEG Findings in Patients Undergoing Anesthesia

With progress in anesthesiology and surgery, the number of old patients undergoing operations has increased. Accordingly, in the EEG studies of 1000 patients, 750 patients were 20–50 and 250 over 70 years of age. This corresponds fairly well with today's standard distribution of patients undergoing surgery. Table 9, summarizing the distribution of EEG background activity in the 1000 patients, including geriatric patients, thus differs from Table 4, which deals with the average population. Overall, there is a smaller proportion of alpha EEGs (66%) and an increased percentage of irregular EEGs and 7–8 Hz variants (12% and 6%, respectively) in the 1000 patients than in the normal population. The influence of hospitalization is only seen in a higher percentage of beta activity, this has to be regarded as drug-induced.

Table 9. Distribution of basic EEG-types and variants in 1000 patients aged 20–90 years – combined classification according to JUNG (1953), OBRIST [13], and KUBICKI and HOELLER [8]

EEG-type	Distribution (%)
Alpha/alpha-beta	37/29 (=66)
Beta	3
Irregular	12
Low-voltage	11
7–8-Hz variant	6
Delta/theta	2

References

A. Textbooks and Summaries

Chatrian GE, Lairy GC (1976) The EEG of the waking adult. In: Remond A (ed) Handbook of electroencephalography and clinical neurophysiology, vol 6A. Elsevier, Amsterdam
Cohn R (1949) Clinical Electroencephalography. McGraw-Hill, New York
Dumermuth G (1976) Elektroenzephalographie im Kindesalter. 3rd edn. Thieme, Stuttgart
Gibbs FA, Gibbs EL (1950) Atlas of Electroencephalography, vols I–III. Addison-Wesley, Reading, MA
Gibbs FA, Gibbs EL (1971) Elektroenzephalographie. Fischer, Stuttgart
Jung R (1953) Neurophysiologische Untersuchungsmethoden: II. Das Elektroenzephalogramm (EEG). Springer, Berlin Göttingen Heidelberg, pp 1216–1325 (Handbuch der inneren Medizin, vol 5/1)
Künkel H (1980) Die Grundphänomene des Elektroenzephalogramms. In: Gruhle HW, Jung R, Mayer-Gross W, Müller A (eds) Handbuch: Psychiatrie der Gegenwart I/2, 2nd edn. Springer, Berlin Heidelberg New York, pp 116–196
Kugler J (1981) Elektroenzephalographie in Klinik und Praxis, 3rd edn. Thieme, Stuttgart

Obrist WD, Busse EW (1965) Electroencephalogram in Old Age. In: Wilson WP (ed) Applications of electroencepnalography in psychiatry. Duke University Press. Durham, pp 185–205

Simon O (1977) Das Elektroenzephalogramm. Urban und Schwarzenberg, Munich

B. Papers

1. Adams A (1959) Studies of flat electroencephalogram in man. Electroencephalogr Clin Neurophysiol 11:34–41
2. Brazier MAB, Finesinger JE (1944) Characteristics of the normal electroencephalogram: I. A study of the occipital cortical potentials in 500 normal adults. J Clin Invest 23:303–311
3. Drechsler F (1978) Quantitative analysis of neurophysiological processes of aging CNS.J Neurol 218:197–213
4. Finley K (1944) On the occurrence of rapid frequency potential changes in the human electroencephalogram. Am J Psychiatry 101:194–200
5. Gallais P, Collomb H, Milletto G, Cardaire G, Blanc-Garing J (1957) Confrontation entre les données de l'électroencéphalogramme et des examens psychologiques chez 522 sujets repartis en trois groupes différentes: II. Confrontations des données de l'électroencéphalogramme et de l'examens psychologique chez 113 jeunes soldats. In: Fischgold H, Gastaut H (eds) Conditionnement et réactivité en électroencéphalographique. Electroencephalogr Clin Neurophysiol [Suppl] 6:294–303
6. Gibbs FA, Gibbs EL, Lennox WG (1943) Electroencephalographic classification of epileptic patients and control subjects. Arch Neurol 50:111–128
7. Jasper H, Penfield W (1949) Electrocorticograms in man: Effect of voluntary movement upon the electrical activity of the precentral gyrus. Arch Psychiatr Nervenkr 183:163–174
8. Kubicki S, Hoeller L (1980) Systematische Einteilung der EEG-Grundrhythmen und -Normvarianten. Das EEG-Labor 2:32–52
9. Kugler J, Oswald WD, Herzfeld U, Seuser R, Pingel J, Welzel D (1978) Langzeittherapie altersbedingter Insuffizienzerscheinungen des Gehirns. Dtsch Med Wochenschr 103:456–462
10. Martelli G (1964) Il reperto elettroencefalografico nelle psicosi arteriopatiche e nelle demenze senili. G Psichiat Neuropat 92:115–146
11. Mundy-Castle AC (1951) Theta and Beta rhythm in the electroencephalogram of normal adult. Electroencephalogr Clin Neurophysiol 3:477–486
12. Mundy-Castle AC, McKiever BL, Prinsloo TA (1953) A comparative study of the electroencephalograms of normal Africans and Europeans of southern Africa. Electroencephalogr Clin Neurophysiol 5:533–543
13. Obrist WD (1954) The electroencephalogram of normal aged adults. Electroencephalogr Clin Neurophysiol 6:235–244
14. Picard P, Navarranne P, Lavourer P, Grousset G, Jest C (1957) Confrontations des données de l'électroencéphalogramme et des examens psychologiques chez 522 sujets repartis en trois groupes différentes. III. Confrontations des données de l'électroencéphalogramme et de l'examens psychologique chez 309 candidats pilotes à l'aéronautique. In: Fischgold H, Gastaut H (eds) Conditionnement et réactivité en électroencéphalographique. Electroencephalogr Clin Neurophysiol [Suppl] 6:304–314
15. Pine I, Pine HM (1953) Clinical analysis of patients with low-voltage-EEG. J Nerv Ment Dis 117:191–198
16. Richter K (1960) Über Anlagefaktoren im EEG. Fortschr Neurol Psychiatr 28:332
17. Roger A, Bert J (1959) Etude des corrélations entre les differentes variables EEG. Rev Neurol (Paris) 101:334–360
18. Vogel F (1962) Untersuchungen zur Genetik des Beta-Wellen-EEG beim Menschen. Dtsch Z Nervenheilk 184:137–173
19. Vogel F (1963) Genetische Aspekte des Elektroenzephalogramms. Dtsch Med Wochenschr 88:1748–1759
20. Vogel F, Fujiya Y (1969) The incidence of some inherited EEG variants in normal Japanese and German males. Hum Genet 7:38–42
21. Vogel F, Götze W (1959) Familienuntersuchungen zur Genetik des normalen Elektroenzephalogramms. Dtsch Z Nervenheilk 178:668–700

B. Electroencephalographic Patterns Induced by Various Anesthetics and Perioperative Influences

I. Premedication

CONTENTS

The use of medicaments to prepare patients for anesthesia and surgery goes back to Schneiderlinn [23], who was the first to use a combination of morphine and scopolamine in 1900. His aims were to reduce fear and adrenaline-induced side effects in the patient, to ease anesthetic induction and maintenance, and incur vagal suppression. These aims still apply today. Adequate and timely premedication for patients undergoing surgery is an important part of routine anesthesiology. The large number of substances available with various sedative, analgetic, antiemetic, and vagolytic effects and antihistaminic properties enable the objectives to be attained fairly closely. These objectives are determined by the patient's psychic and physical condition, by the duration and extent of the planned surgery, and by long-term effects that are occassionally desired after the completion of surgery. Even with exact weight-related doses, it is difficult to predict the clinical effect of a drug on the individual patient due to the different effects with respect to the drug's extent and mode of action. Therefore, in routine anesthesiology, a certain standardization of premedication has been established. This does not always, however, suit the particular situation of the patient.

A great number of recently performed studies on the use of mixed substances in premedication suggest that there is uncertainty as to the most suitable means of performing this anesthesiological procedure. The exact assessment of efficient premedication – especially with regard to particular effects – is further complicated by subsequent anesthesia and surgery. Such assessments are frequently dismissed because of a lack of postoperative anesthesiological observation. Therefore, a clinical evaluation of the usual premedicative measures should be as precise as possible and be combined with an objective method of studying the mode of action and strength of the substances used. Since in premedication it is largely the effects of drugs on the nervous system that come under consideration, the EEG appears suitable as an objective method of evaluation. Here, general statements about the functional changes in the brain caused by a particular substance as well as about the occurrence and degree of the changes in the individual case are required. From individual deviations observed when assessing the initial parameters, conclusions can be drawn about the adequate dosage or selection of drugs for patients with similar patterns.

In the light of such considerations, the clinical and EEG effects analyzed in seven substances used today in premedication will be described below. This will be based on our own uniform and comparative studies in a total of 109 patients as well as on appropriate reports in the literature [1, 3, 7, 20]. Because of marked age-dependent EEG differences, our own studies involved two groups of patients aged 20–50 years and over 70 years. After recording a 5-min resting EEG, the premedicant was injected intravenously. The EEG recordings were usually performed for 20 min, in particular cases for one or even several hours. In addition to the test substance, all patients received 0.25 mg atropine for forensic reasons.

Single Substances

Atropine (Atropin)

Pharmacological Mode of Action and Clinical Application. Of primary anesthesiological significance is the anticholinergic effect of atropine on salivation and the vagosuppressive effect of atropine on the heart. Atropine is today widely administered as a routine premedicant. It mainly counteracts salivation or marked bradycardia, which might occur as a result of sympathicolysis from the application of the anesthetic or through direct surgical manipulation. The usual dose of 0.05–0.5 mg in the child and 0.5–1.0 mg in the adult is as a rule sufficient for premedication; at the onset of bradycardia, the dose of atropine has to be repeated as required. The central action of atropine, an anticholinergic effect on the neurons of the central nervous system, is of no clinical significance. The depression of the respiratory center by morphine can be relieved by atropine through stimulation of the central nervous system. In higher doses, atropine acts as a centrally depressive substance.

EEG Findings

Our findings. These were derived from eight patients 20–50 years of age and eight patients over 70 years; the dose given was 0.5 mg atropine i.v. [15]

Table 1. EEG-changes after intravenous atropine administration

Frequency range	EEG power suppression after application of drug	Distribution (%)	Degree of change
Alpha	↑–↑↓–↓[a]	6–13–25	ca. 5–10 μV
Beta	–↓[a]	6	
Delta	↑[b]	62	ca. 10–20 μV
Theta	↑[b]		
	No change		

[a] ↑ increase, ↓ decrease, ↑↓ varying increase and decrease
[b] Especially in the lower age-group

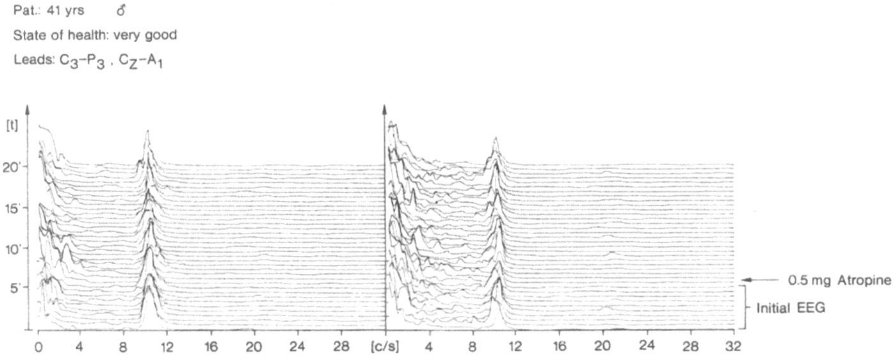

Fig. 1. EEG spectral analysis prior to and after i.v. injection of 0.5 mg atropine. The dominant initial alpha band is slightly affected. The occasional visible suppression of alpha activity is probably due to vigilance changes.
Recording conditions: time constant (TC), 0.3 s; filter, 70 Hz; calibration, 50 μV = 7 mm; paper speed, 30 mm/s; FFT, 30-s periods

(Fig. 1; Table 1). The observed EEG changes were always slight and inconsistent. They frequently failed to appear at all, especially in the upper age-group.

Review of Other Findings. With regard to atropine application in psychiatric disease, an increase in electric power in the beta and theta range and a decrease in the alpha range have been described with higher atropine doses [6].

Clinical Evaluation. Apart from the desired diminution of salivary secretion, no effects, especially central nervous effects, were observed in general. The average pulse rate after atropine application was not raised significantly. Individually, moderate rises in the pulse rate may be found, especially in younger subjects.

Interpretation of EEG Changes and Clinical Findings. In accordance with the minimal effects of atropine – at the given dose – a significant EEG change at the neurons of the CNS was not expected and did not occur in our own patients. Therefore, no conclusions regarding adequate premedication with atropine can be drawn from the EEG pattern. The observed changes, especially in the alpha range, may be due to physiological changes in alertness or may be indistinguishable from them. The small increase in activity in the delta and theta ranges, occurring especially in the lower age-group, can be interpreted as a manifestation of calmness, which could be observed clinically and somewhat independent of the atropine application.

Diazepam (Valium)

Pharmacological Mode of Action and Clinical Application. Diazepam has found a wide use as an effective sedative. Anxiolytic central sites of action are the cortex and subcortical structures, especially the limbic system. Diazepam leads to an indifferent psychic condition with sedation and loss of mental depth; at higher doses, this is accompanied by tiredness and somnolence. The effects of alcohol and hypnotics are potentiated. Confusion – combined with motor restlessness – is observed as the most striking side effects. With continuous use, psychophysical dependence results with withdrawal reactions upon discontinuation of therapy. Further indications or diazepam use are seen particularly in psychotherapy and neurology; here, diazepam is used alone, even in higher doses, or in combination with other substances. In anesthesiology, diazepam is also used frequently for several indications – for sedation in premedication and intensive care, as a combination anesthetic in anesthesia, and as an anticonvulsive in the treatment of eclampsia and tetanus.

EEG Findings

Our Findings. These were derived from eight patients aged 20–50 years and eight patients over 70 years; 20 mg Valium i.v. [18] (Fig. 2; Table 2). Independent of the initial EEG, an increase in power can be seen as a characteristic change in the beta band between 12.5 and 22 Hz. There was also a reduction in alpha activity and an additional frequency shift of 0.5–1 Hz toward the lower range. Transient increases in power in the delta/theta band between 3 and 3 Hz were also found. Individually, the degree of these changes varied markedly (also corresponding with the degree of sedation, see below).

Review of Other Findings. The described increase in beta activity is known from EEG studies in patients undergoing psychiatric or neurological treatment. Preexisting abnormal frequency patterns become less noticeable un-

Table 2. EEG-changes after intravenous diazepam administration

Frequency range	Type of power change after injection of drug	Distribution (%)	Intensity of changes
Alpha	a) ↓ b) ↓ DF	a) 100 b) 80	a) 10–50 μV b) 1– 2 Hz
Beta	↑	100	(at 12.5–22 Hz) (20–50 μV)
Delta	↑	30	50–100 μV
Theta	↑		

Abbreviation: DF, dominant frequency in the initial EEG

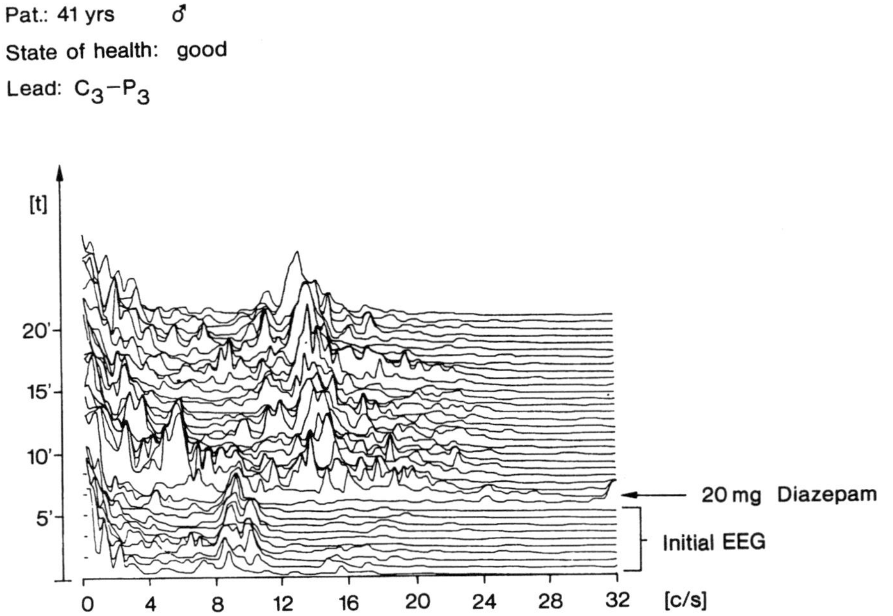

Fig. 2. Typical diazepam-induced changes in EEG spectral analysis with clinically excellent premidication. First, the initial alpha EEG (9–10 Hz) is replaced by a high-voltage irregular delta/theta and beta activity. Subsequently, activation of the beta band predominates, which is characteristic for the diazepams.
Recording conditions: TC, 0.3 s; filter, 70 Hz, calibration, 50 μV = 7 mm; paper speed, 30 mm/s; FFT, 30-s periods

der diazepam. These findings are interpreted as suggestive of a subcortical influence of diazepam in seizures and certain states of restlessness [10, 12, 25, 26].

Clinical Evaluation. After a few minutes, the intravenous application of 20 mg diazepam leads to marked psychic indifference with feelings of well-being, although at the same time there is a tendency for childish – frivolous behavior. In addition, there are individually varying levels of sedation and, especially in the upper age-group, the diazepam leads to mild or deep stages of sleep. Despite psychic indifference, in approximately one-third of patients, confusion and motor restlessness are observed as side effects. These side effects occur especially in combination with more pronounced sedation or slight stages of sleep. In the age-group of over 70 years, a mild yet significant decrease in blood pressure occurs.

Interpretation of EEG Changes and Clinical Findings. The increase of activity in the beta range, which is especially striking and always present, is probably related to the major effect of diazepam, that of inducing psychic indifference. The degree of EEG changes and psychic indifference vary individually but are related. With a decrease in alpha activity, more pronounced sedation can be found; side effects are not seen. Clinically undesirable central side effects are observed in the case of slight changes in the alpha activity and only slight increases in the beta activity. The occurrence of frequencies between 2 and 7 Hz coincides with tiredness and induction of sleep. A correlation can also be found between the degree of increased amplitude and depth of sleep. Thus, the EEG not only indicates the general mode of action of diazepam, from the degree of the characteristic changes, it also allows individual conclusions as to its central effects to be made.

Triflupromazine (Psyquil)

Pharmacological Mode of Action and Clinical Application. Triflupromazine belongs to the group of phenothiazine derivatives. It has an inhibitory action on functions of the tectoreticular system, the thalamus, the area postrema, and the vomiting center. Peripherally, it has mild anticholinergic and adrenolytic effects. Clinically, the central influence leads to calming and, occasionally, to the induction of sleep. The effects of analgetics, sedatives, hypnotics, and narcotics are potentiated by this drug. Triflupromazine, although it belongs to the tranquilizers, acts neuroleptically in higher doses. In anesthesiology, the substance is used as a tranquilizer and in particular also for central suppression of vomiting; it is occasionally used for premedication, though more frequently for postoperative treatment [24].

EEG Findings

Our Findings. These were derived from 12 patients aged 20–50 years; the dose given was 10 mg triflupromazine i.v. [19] (Fig. 3; Table 3). The EEG changes with triflupromazine are pronounced to different degrees, and occasionally show different characteristics according to the weight-related

Table 3. EEG-changes after intravenous triflupromazine administration

Frequency range	Type of power change after injection of drug	Distribution (%)	Intensity of changes
Alpha	a) ↓ (2– 7 min) b) ↓ (7–20 min) c) Decrease in frequency	56	10–40 µV 1/2–1 Hz
Beta	↑	12.5	(12.5–22 Hz) 5– 20 µV
Delta Theta	↑ ↑	80	50–100 µV

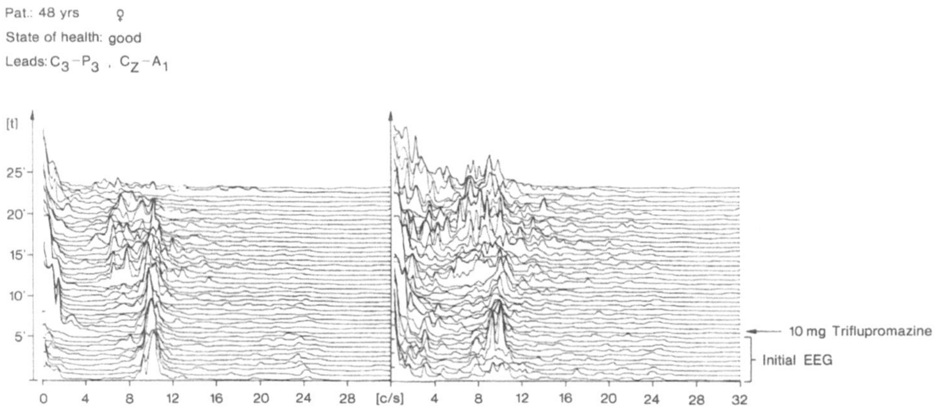

Fig. 3. EEG spectral analysis prior to and after i.v. injection of 10 mg triflupromazine. The initial EEG is a normal alpha EEG with a dominant frequency of 10 Hz. After injection of triflupromazine there is a brief suppression of the initial activity, after which the alpha band is reactivated. Additionally, as a manifestation of the good sedation, there is a clear increase in activity in the theta range.
Recording conditions: TC, 0.3 s; filter, 70 Hz; calibration, 50 µV = 7 mm; paper speed, 30 mm/s; FFT, 30-s periods

dose. Most frequently, within 15 min, medium to pronounced delta and theta activities develop together with changes in the alpha range. Over a brief decrease in alpha activity, lasting approximately 7 min, an increase in activity and deceleration of frequency of 0.5–1 Hz develop. Occasionally, with higher doses (0.15–0.18 mg/kg), additional beta activities of 12.5–22 Hz occur – approximately 10 min after injection. An alpha frequency deceleration alone or no change at all is seen in a few cases.

Review of Other Findings. In the classification of psychotropic substances of Itil [5], neuroleptics show psychophysically calming effects evident of an increase in delta/theta power.

Clinical Evaluation. Triflupromazine leads to slowly increasing sedation in addition to its entiemetic effect, which was not tested here. Subjects reported that they experienced tiredness and the desire to sleep. In some cases, no change in the physical and emotional condition was noticed both subjectively and objectively. Blood pressure dropped slightly; the pulse rate did not change significantly; shortly after injection, a transient tachycardia occcasionally occurred.

Interpretation of EEG Changes and Clinical Findings. Generally, the EEG changes correlate clinically with sedative and sleep-inducing effects. The application of 10 mg triflupromazine i.v. leads to variable EEG changes, depending on the individual weight-related dose. Combinations of changes in several frequency ranges occur in each case. The increase in power in the delta/theta range parallels sedation and sleep. The findings of reduced or increased activity in the alpha frequency band are clinically associated with tranquilization or increased sense of well-being. With higher doses beta activity occurs along with phases of sleep. An assessment of the clinical premedicative effect of triflupromazine is possible both generally and in the individual case.

Promethazine (Atosil)

Pharmacological Mode of Action and Clinical Application. Promethazine is a phenothiazine derivative with central and peripheral sites of action. It is classified as a neuroleptic. The reticular formation, the hypothalamus, and possibly the limbic system (the tectoreticular system, the thalamus, and the area postrema) are regarded as central sites of action. The central effect is characterized by psychomotor suppression, sleep-inducing effects, potentiation of sedative drugs, inhibition of the vomiting center, and influences on central thermoregulation. Peripherally, there is an antiadrenergic effect, due to incomplete alpha-receptor blockage, and antihistaminic properties, due to competitive blocking of histamine receptors in the tissue. Mainly because of its central effects and antihistaminic properties, promethazine is used for premedication, the potentiation of anesthesia, and postoperative analgetic therapy. Its application is also seen in supportive sedation and temperature-decreasing measures in intensive care. In anesthesiology, promethazine is injected in single doses of 25–50 mg, intravenously or intramuscularly.

EEG Findings

Our Findings. These were derived from nine patients aged 20–50 years and eight patients over 70 years; the dose given was 50 mg promethazine i.v. [14] (Fig. 4; Table 4). The changes with promethazine recorded by EEG

Table 4. EEG changes after intravenous promethazine administration

Frequency range	Type of power change after injection of drug	Distribution (%)	Intensity of changes
Alpha	a) At alpha initial EEG (88%)	100	10 –40 μV
	b) ↓ DF	26	0.5– 1.5 Hz
Beta	a) ↓ Of higher ranges with irregular initial EEG (12%)	100	(16–32 Hz)
Delta Theta	↑ ↑ } Independent of initial EEG	100	20–70 μV

Abbreviation: DF, dominant frequency in the initial EEG

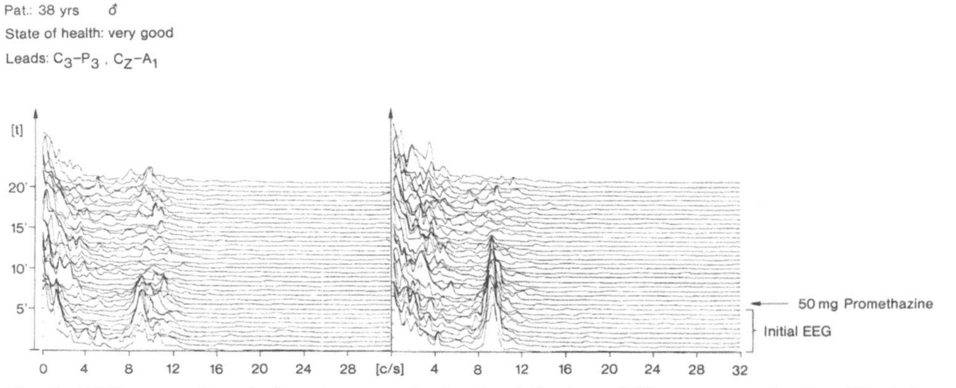

Fig. 4. EEG spectral analysis prior to and after i.v. injection of 50 mg promethazine. The initial EEG demonstrates alpha activity with a dominant frequency of 9 Hz. After promethazine, there is a marked sustained decrease in the alpha band and an increase in the delta and theta ranges. Clinically, the changes were accompanied by a good sedative effect.
Recording conditions: TC, 0.3 s; filter, 70 Hz; calibration, 50 μV = 7 mm; paper speed; 30 mm/s; FFT, 30-s periods

were characterized by the suppression of alpha activity, with individual amplitude decreases of between 20% and 70%. There were also frequency decreases of 0.5–1.5 Hz (with an initial pattern of alpha or alpha/beta) or reductions in the beta fraction in the case of an irregular initial EEG and an increase in delta/theta activity.

Review of Other Findings. After application of various substances of the phenothiazine group, EEG changes have mainly been studied during psychiatric long-term treatment [4]. The drugs are used because of their antipsychotic and sedative effects. It has been shown that a slow intravenous application of promethazine (approximately 1 mg/kg over 10 min) initially diminishes the voltage of the background activity; it leads also to the occurrence of fast and slow waves to a lesser degree. Occasionally, stages of sleep are observed with the simultaneous occurrence of beta-spindles between 18 and 32 Hz. These findings, with the exception of the occurrence of beta spindles, are in agreement with our own observations. Beta spindles, as transient graphoelements, cannot be determined precisely by spectral analysis and thus probably were unrecognized in our own patients.

Clinical Evaluation. Promethazine causes mild to moderate sedation with occasional light periods of sleep in the manner of application and dosage mentioned above. The extent of clinically premedicative effects varies individually but is generally more pronounced in older patients. Frequently, in 62% of patients, despite central sedation, severe degrees of psychomotor restlessness occur. Subjectively, these are unpleasant. Blood pressure and pulse rate show minute yet significant increases.

Interpretation of EEG Changes and Clinical Findings. The divergent clinical effects of promethazine in premedication, including sedation as well as psychomotor restlessness, are manifest in the EEG changes. Decrease of power in the alpha band, or in the beta band with an irregular initial EEG, parallel clinical calming effects. An increase in delta/theta activity is always an expression of sleep-triggering effects. A slight decrease in power in the alpha band, despite an increase in the low frequencies, reflects the clinical picture of psychomotor restlessness. Clinical effects and EEG findings can be related.

Pethidine (Dolantin)

Pharmacological Mode of Action and Clinical Application. Pethidine is a synthetic opiate. In premedication, its specifically analgetic effect has been used for preoperative psychic and physical calming since 1939. This effect combines with sedative and dysphoretic or slightly euphoretic properties [21, 22]. Central major sites of action are the cortex and diencephalic centres. Apart from the desired central effects, as with all opiatelike analgetics, the respiratory center is depressed in a dose-dependent fashion. Peripherally, pethidine leads to a decrease in tone of smooth muscle, with occasional reductions in blood pressure. Side effects, such as perspiration, dizziness, and tingling sensations in the extremities are the combined effects of a lowered blood pressure and restricted respiratory function. A single dose in children and adults is about 1 mg/kg.

EEG Findings

Our Findings. These were derived from eight patients aged 20–50 years and eight patients over 70 years; the dose given was 100 mg Dolantin i.v. [13]

Table 5. EEG changes after intravenous pethidine administration

Frequency range	Type of power change after injection of drug	Distribution (%)	Intensity of changes
Alpha	a) ↓ b) ↓↑ At initial alpha EEG (69%)	a) 90 b. 10	10–70 μV
Beta	At irregular initial EEG (31%)	100	(16–28 Hz) 20 μV
Delta Theta	↑ Independent of ↑ the initial EEG	} 69	} 5–40 μV

(Fig. 5; Table 5). Pethidine leads to unequivocal and largely undirectional changes in the EEG. The individual range of variation in the extent of reactions is wide. A characteristic feature is the decrease in alpha activity in all patients with an alpha or alpha/beta initial EEG: It ranges from a slight to considerable decrease in amplitude to the complete disappearance of the alpha or alpha/beta activity. In older patients with an irregular initial EEG of preexisting frequencies in the delta to midbeta ranges, the frequency band is narrowed at the expense of the higher ranges. This means that in the beta band a decrease in activity occurs, the extent of which varies between individuals. Delta and theta frequencies are exclusively activated in patients with an initial alpha activity or primary low-voltage EEG.

Clinical Evaluation. Apart from analgetic effects, which during anesthesia provide basic analgesia. pethidine at the applied intravenous dose, leads to good or excellent sedation with light to moderate sleep after 2–3 min. The full effect is maintained for 10–15 min in young patients and for 30–60 min in older ones. Hangovers always occur. In 15% of patients, mild side effects in the form of psychomotor restlessness are present. Significant changes in blood pressure and pulse rate did not occur in our own patients.

Interpretation of EEG changes and clinical findings. Central reactions, which are always detectable in the EEG, directly correlate in their severity with the extent of the observed premedicative effects. The decrease in power between 8 and 12 Hz (in patients with a normal initial EEG) or the reduction in beta activity (in patients with an irregular initial EEG) is the manifestation of the indifferent, calming psychophysical behavior; the increase of activity in the delta-theta range induces sleep. Psychomotor restlessness is accompanied in the EEG by a considerable increase in the low frequencies and a mild decrease in alpha or beta. The central effects of pethidine correlate well with the observed EEG changes.

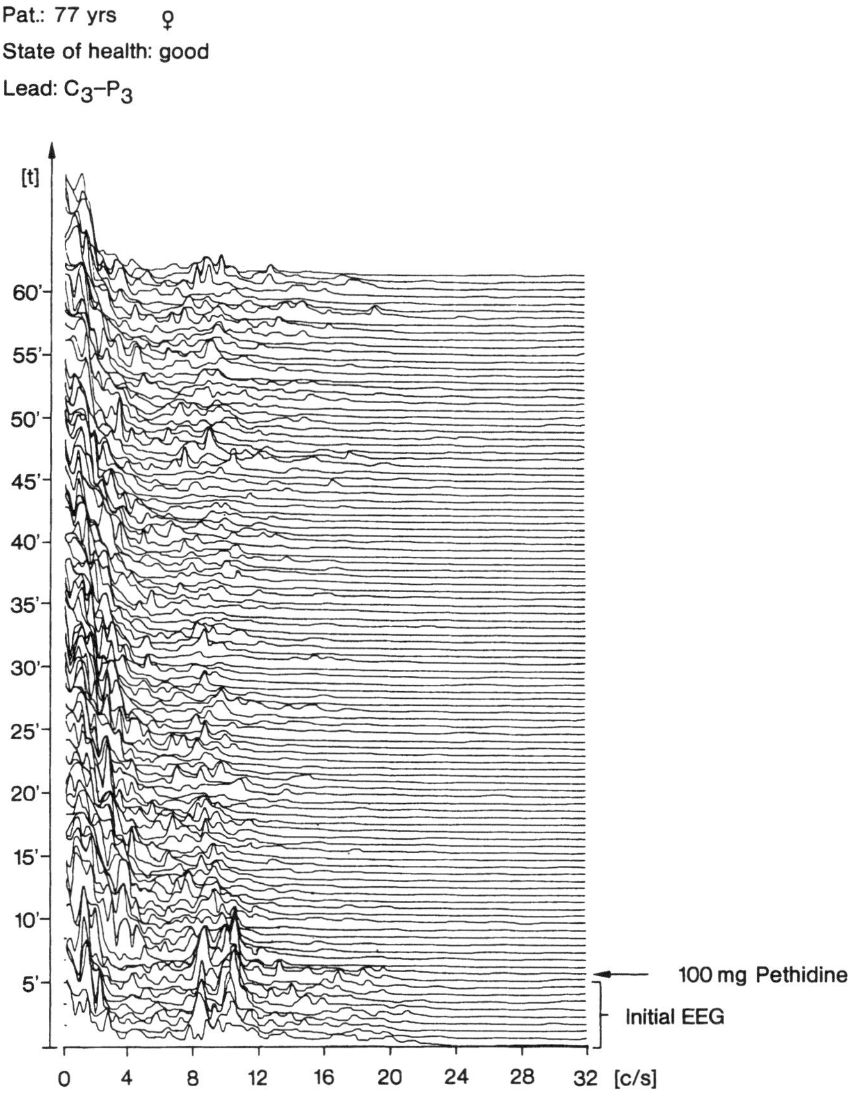

Fig. 5. EEG spectral analysis prior to and after i.v. injection of 100 mg pethidine. Initially, there is an alpha EEG of 8–12 Hz. The injection of pethidine leads to a reduction in voltage and frequency in the alpha band and to a marked increase of activity in the delta range. The changes, corresponding with the good sedative effect, remain constant during the entire observation period. Recording conditions: TC, 0.3 s; filter; 70 Hz; calibration, 50 μV = 7 mm; paper speed, 30 mm/s; FFT 30-s periods

Compound Preparations

Pethidine and Promethazine (Dolantin and Atosil)

Pharmacological Mode of Action and Clinical Application. Pharmacologically, the combination of these substances is especially useful for the purposes of premedication. The central sedative effect with psychophysical calming of the patient and a shielding effect against external and internal stimuli is obtained with pethidine and promethazine. These substances partly act additively and partly potentiate each other. They lead to a long sustained basic sedation with advantages for anesthesia and the postoperative phase. This combination has proven clinically successful for more than three decades.

EEG Findings

Our Findings. These were derived from eight patients aged 20–50 years and eight patients over 70 years; the dose given was 50 mg Dolantin and 25 mg Atosil [16] (Fig. 6; Table 6). A very constant and pronounced decrease in activity in the alpha or beta range, depending on the initial EEG, and an increase of activity in the delta and theta range, which is independent of the initial EEG, occur after application. Compared with the general unidirectional EEG changes after pethidine or promethazine alone, the combination of both at half the dose always leads to therapeutic success, with increased effectivity and fewer side effects.

Clinical Evaluation. In all patients, a feeling of well-being and increasing tiredness occur within 1–2 min. After approximately 5 min, deeper sedation and sleep occur. The depth of sleep is less pronounced in younger patients than in geriatric ones. The period of effective action is 1–3 h. Central side effects, if any, are mild and rarely seen. Blood pressure and pulse rate show a slight yet significant increase.

Interpretation of EEG Changes and Clinical Findings. With this premedication, the consistency and concordance of the psychic changes with those of the EEG are characteristic and indicative of central sedation. The long sustained and marked decrease in the dominant background activity in the EEG correlates with the feeling of well-being and considerable sedation; the degree and duration of increased activity between 2 and 6 Hz are equivalent to the depth and duration of the sleep phases. Thus, from the EEG, the effect of the combination pethidine/promethazine and its duration can be determined.

Table 6. EEG changes after intravenous pethidine-promethazine administration

Frequency range	Type of power change after injection of drug	Distribution (%)	Intensity of changes
Alpha	↓ At initial alpha EEG (87.5%)	100	20–70 μV
Beta	↓ Of higher ranges at irregular initial EEG (12.5%)		(16–28 Hz) 30 μV
Delta	↑ Independent of initial EEG		50–100 μV
Theta	↓ Independent of initial EEG		

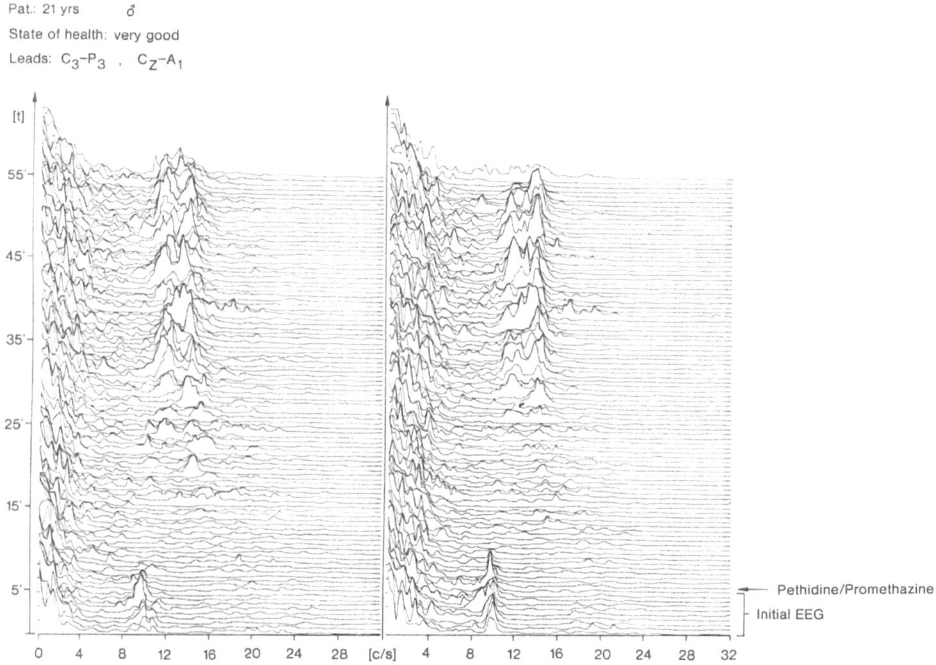

Fig. 6. EEG spectral analysis prior to and after 50 mg i.v. pethidine and 25 mg i.v. promethazine. The initial alpha EEG with an average frequency of 9 Hz is suppressed for 20 min, it returns with a markedly increased voltage and frequency (11–14 Hz). Additionally, there is a slight increase of activity in the theta range.
Recording conditions: TC, 0.3 s; filter, 70 Hz; calibration 50 μV = 7 mm; paper speed, 30 mm/s; FFT, 30-s periods

Droperidol and Fentanyl (Mixture 50:1 = Thalamonal)

Pharmacological Mode of Action and Clinical Application. Thalamonal is a compound preparation containing dehydrobenzperidol (DHB) and fentanyl. These are the two active substances in the classic neuroleptanalgesia

(1 ml = 2.5 mg DHB and 0.05 mg fentanyl). Thalamonal inhibits the reticular formation, thereby suppressing peripheral influences on the cerebrum. The ventricular area is also inhibited. The most prominent peripheral effect is an alpha-receptor blockage at the blood vessels, with dilatation of the lumen and decrease in blood pressure – especially in hypovolemia. Thalamonal was first used for premedication of neuroleptanalgesia in doses of 1–2 ml i.m. Today, it is used widely and independent of the planned form of anesthesia: Application is in premedication, intraoperative conduction anesthesia, and sedation in intensive care, the latter involving intravenous use. Thalamonal largely meets the demands for drug premedication in surgery, which are psychic calming, decreased response to external stimuli, and prevention of vomiting. The Thalamonal effect protects the patient against the disturbing influences of preparation in the operating room. Since, however, there is only an indirect effect on the cortex, states of central excitation are not suppressed. Occasionally, this leads to the phenomenon of a patient complaining of severe inner restlessness with physical stiffening spreading outward; the physician has the impression that the patient is undergoing extensive psychic and physical suppression.

EEG Findings

Our Findings. These were derived from eight patients aged 20–50 years and eight patients over 70 years; the dose given was 2 ml Thalamonal i.v. [17] (Fig. 7; Table 7). Alpha activity was enhanced and stabilized, i.e., the voltage was slightly increased and could not be modulated by external influences. With a low-voltage initial EEG, there was activation of a low-voltage alpha rhythm; with preexisting alpha activity, it was 0.5–1 Hz lower than in the initial findings. This might suggest an influence on cerebral metabolism, analogous to similar age-induced changes [9]. With an irregular initial EEG, the frequency range in the higher beta range was diminished by 2–6 Hz or the activity in this range was markedly suppressed. There was also always an increase in activity between 2 and 4 Hz. The observed changes were mild but constant.

Review of Other Findings. Kubicki and Zaduk [8] studied the effects of droperidol and fentanyl on the EEG. This was done during neuroleptanalgesia with a correspondingly higher dose. The so-called stabilization of alpha activity was found to be a characteristic change, which means that the alpha rhythm was unchanged by external influences and there was a synchronization of the EEG frequencies. Linde et al. [11], who studied the EEG effects of Thalamonal in premedication, found no significant changes. These different results – Kubicki and Zaduk [8] noted marked EEG changes, our own findings indicated moderate EEG changes, Linde et al. [11] found no changes – must be explained by the different doses and modes of application (compare with Chapter BIV).

Table 7. EEG changes after intravenous Thalamonal administration

Frequency range	Type of power change after injection of drug	Distribution (%)	Intensity of changes
Alpha	At initial alpha EEG (62%) a) ↑ b) ↓ DF With low voltage initial EEG (19%) c) ↑	100	10 μV 0.5–1 Hz 5–15 μV
Beta	↓ With irregular initial EEG (19%)	100	10–20 μV
Delta	↑ Independent of initial EEG	100	20–50 μV
Theta	–	–	–

Abbreviation: DF, dominant frequency in the initial EEG

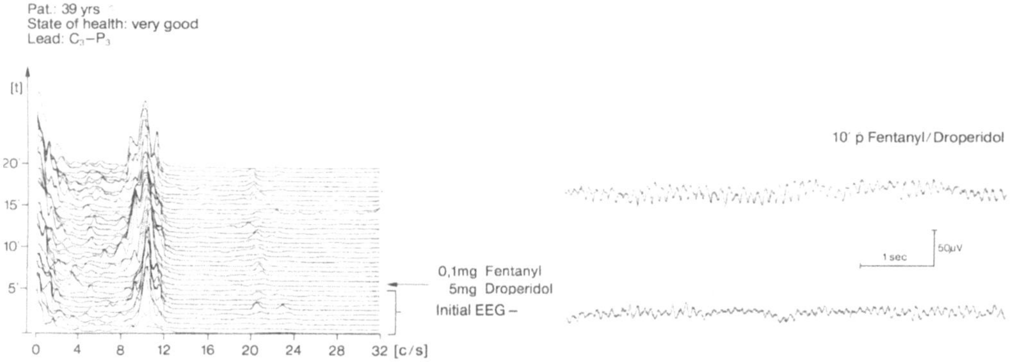

Fig. 7. EEG spectral analysis and conventional EEG prior to and after i.v. 0.1 mg Fentanyl and 5 mg Droperidol. The initial EEG is an alpha EEG with a dominant frequency of 10.5 Hz. After injection there is an increase in voltage and moderate slowing of frequency (0.25–0.5 Hz). Clinically, the changes were accompanied by good sedation.
Recording conditions: TC, 0.3 s; filter, 70 Hz; calibration 50 μV=7 mm; paper speed, 30 mm/s; FFT, 30-s periods

Clinical Evaluation. Thalamonal leads to psychic and physical calming with slight degrees of sedation and occasionally tiredness and brief periods of sleep. Compared with the typical and constant sedation after, for example, premedication with pethidine/promethazine, patients are conscious and responsive after Thalamonal treatment. In the upper age-group, the premedicative effect is more pronounced. Side effects as mentioned above (central excitation) generally are very rare and did not occur in the patients evaluated here. The most frequent side effect is a slight decrease in blood pressure. In younger patients, this is due to the peripherally vasodilating mech-

Table 8. Summary of major changes in the EEG and clinical pattern with various forms of premedication ($n = 109$)

Substance and dose	EEG	Clinical response[a]	
		Premedicative effects	Side effects
Atropine, 0.5 mg i.v.	(Mild changes at tested doses) Slight power increase or decrease in alpha activity Slight power decrease in beta activity Slight power increase in delta/theta activity	No sedation Sufficient reduction of salivation	Central – none Otherwise – mild increase in pulse rate
Valium, 20 mg i.v.	Increase in activity in beta range (12.5–22 Hz) Occasional reduction or extinction of alpha activity (with initial alpha EEG) or accentuation of low-frequency beta range (12.5–22 Hz) (with irregular initial EEG) Increase of activity in theta range	Psychic indifference Sedation – + to + + + Sleep – none to + + +	Central – mild degrees of psychomotor restlessness, ca. 35%. Otherwise slight decrease in blood pressure, mainly in geriatric patients
Triflupromazine, 10 mg i.v.	After transient alpha reduction, slight increase in activity and decrease in frequency at 0.5–1 Hz In some cases, increase of activity in beta range (12.5–22 Hz) Increase of activity in delta/theta range	Euphoria, sedation – + to + + + Sleep – (+) to +	Central – none or rarely restlessness. Otherwise – mild decrease in blood pressure, occasional transient tachycardia
Promethazine, 50 mg i.v.	Reduction of alpha activity < 50% (with initial alpha EEG), or Narrowing of basic frequency power < 50% in higher frequency range (with irregular initial EEG) Increase of activity in delta/theta range	Sedation – + to + + Sleep (+)	Central – severe degrees of psychomotor restlessness, ca. 62%
Pethidine, 100 mg i.v.	Reduction of alpha activity > 50% (with alpha initial EEG), or Narrowing of basic frequency power > 50% in higher frequency range (with irregular initial EEG) Increase of activity in delta/theta range	Sedation – + + to + + + Sleep – + to + +	Central – mild degrees of psychomotor restlessness, ca. 16%

Table 8 (continued)

Substance and dose	EEG	Clinical response[a]	
		Premedicative effects	Side effects
Pethidine + prometha-zine, 50 mg + 25 mg i.v.	Reduction of alpha activity > 50% (with initial alpha EEG), or Narrowing of basic frequency power > 50% in higher frequency range (with irregular initial EEG)	Sedation – + + +	Central – none
	Pronounced increase of activity in delta/theta range	Sleep – + to + + +	
Thalamonal, 2 ml i.v.	Mild accentuation of alpha activity and decrease of alpha frequency by 0.5–1 Hz (at initial alpha EEG), or Mild narrowing of basic frequency power in higher frequency range (with irregular initial EEG)	Psychic shielding Sedation + to + + +	Central – none, or rarely, subjective inner restlessness with catalepsia, otherwise – mild decrease in blood pressure, mainly in young patients
	Slight increase of activity in delta range	Tiredness, occasional sleep – + to + +	

[a] Degree of sedation or sleep: + mild, + + moderate, + + + deep

anism of droperidol and occurs despite prior application of 250 ml colloidal fluid. It occurs less frequently in the upper age-group, probably because of general vascular changes.

Interpretation of EEG Changes and Clinical Findings. The pronounced dominant alpha activity in the EEG parallels the above-mentioned psychic calming of the patient. The maintenance of an alpha activity is in accordance with the described conscious responsiveness and pharmacologically this can be explained by a lack of direct cortical sedation. Increase of activity in the delta range generally indicates induction of sleep. The EEG changes after administration of Thalamonal, due to the unidirectional and characteristic mode of alteration, allow an assessment of the premedicative effect. However, they are less pronounced, and thus more difficult to determine, than in other forms of premedication. The main importance of EEG analysis with Thalamonal is in determining preserved or accentuated and stabilized alpha activity. As evidence of undisturbed regulation of vigilance in shielding against peripheral influences. This was also found by Kubicki and Zaduk [8] in neuroleptanalgesia and was established here in premedication with Thalamonal.

Summary of EEG Findings in Premedication (Table 8)

With regard to their clinical effects, the EEG findings in premedication allow the following conclusions to be made.

1. A Centrally Effective Premedication Leads to Clearly Determinable EEG Changes, Which Depend on the Type of the Initial EEG. According to the central pharmacological action of the substance used in premedication (with the exception of atropine), the EEG changes are found mainly in the alpha range, characterizing the degree of vigilance. In older patients with an irregular initial EG, this corresponds to the predominant beta range. As evidence of deeper sedation, the activities in the delta and theta ranges must be considered and are frequently associated with signs of sleep.

2. The EEG Changes are Frequently Characteristic for Particular Substances Used in Premedication. Depending on the specific cerebral site of action of the substances used in premedication and depending on the degree of their hypnotic or narcotic potential, relatively characteristic EEG changes result. At present, a strict classification of these substances according to particular pharmacological effects and sites of action is possible only to a limited extent. Marked differences, such as the maintenance or accentuation of the alpha range with Thalamonal, however, point to the existence of considerable variations in their modes of action compared with other substances. Knowledge of the EEG changes that are characteristic for a certain substance allows a prediction to be made as to its effects and possible individual variations.

3. There is a Correlation Between the Degree of the EEG Changes and the Severity of Clinical Effects. It can thus be concluded that the degree of EEG changes individually allows a prediction as to the premedicative or sedative effects, which supports the clinical evaluation and can be helpful in the elaboration of special dosage schemes. Differences in the serum levels and clinical effects – frequently determined by the permeation conditions at the blood-brain barrier (e.g., affected by opiates) – can be clarified by EEG control.

4. In the EEG Pattern and Accordingly in the Clinical Findings, Marked Differences Between Various Age-groups and Individual Patients are Seen. With regard to the typical EEG changes with various modes of premedication, individual variations in the type of change are rarely seen, whereas the degree of change is frequently different. If in the older age-groups variants exist in the initial EEG, the EEG pattern typically differs from that in younger age-groups. The reactions of older patients can also differ from those in younger patients in the strength and duration of the drug effects independent of the primary EEG changes. This corresponds with the findings of CASTLEDEN et al. [2], who described a considerably higher cerebral response to nitrazepam in geriatric patients, despite otherwise unaltered pharmacokinetics.

5. The EEG is of Significance in Testing the Substances Used in Premedication. It reveals the mode of action and strength of the drug as well as various types of reaction or probabolities for individual patient-groups to respond to the substance in question. Knowledge of the specific EEG changes is essential for a competent assessment of the EEG during anesthesia.

References

A. Textbooks and Summaries

Benzer H, Frey R, Hügin W, Mayrhofer O (1982) Lehrbuch der Anästhesiologie. Intensivmedizin und Reanimatologie, Springer, Berlin Heidelberg New York

Lee JA, Atkinson RS (1978) Synopsis der Anästhesie. Fischer, Stuttgart

B. Papers

1. Arfel G (1978) Die Kennzeichen des seneszenten EEG. Symposium: Le cerveau agé et ses médicaments, 9–10 Jan 1978, Paris
2. Castleden CM, George CF, Marcer D, Hallet C (1977) Increased sensitivity to nitrazepam in old age. Br Med J 1: 10–12
3. Drechsler F (1978) Quantitative analysis of neurophysiological processes of the aging CNS. J Neurol 218: 197–213
4. Itil T (1961) Elektroenzephalographische Befunde zur Klassifikation neuro- und thymoleptischer Medikamente. Med Exp 5: 347–363
5. Itil T (1969) Definition and classification of neuroleptics from the EEG point of view. In: Bobon DP (ed) Reprints of the interdisciplinary week of neuroleptics. University of Liège, pp 167–169
6. Itil T, Fink M (1966) Anticholinergic drug-induced delirium: Experimental modification, quantitative EEG and behavioral correlations. J Nerv Ment Dis 143: 492–507
7. Karbowski K (1977) Das Alters-EEG. Schweiz Med Wochenschr 107: 1241–1247
8. Kubicki S, Zaduk P (1966) EEG-Veränderungen nach Neuroleptanalgesie. In: Henschel WF (ed) Die Neuroleptanalgesie. Anästhesiologie und Wiederbelebung, vol 9. Springer, Berlin Heidelberg New York, pp 44–49
9. Kugler J, Oswald WD, Herzfeld U, Hus R, Pingel J, Wetzel D (1978) Langzeittherapie altersbedingter Insuffizienzerscheinungen des Gehirns. Dtsch Med Wochenschr 103: 456–462
10. Lieb J, Scabassi R, Wandall P, Buchners R (1974) Comparison of the action of diazepam and phenobarbiturate using EEG derived power spectra obtained from temporal lobe epileptics. Neuropharmacology 13: 769–783
11. Linde L, Seidlitz K, Wallrabe D, Schulz H (1976) Der Einfluß der Prämedikation von Droperidol und Fentanyl auf die Hintergrundaktivität im Elektroenzephalogramm bei zerebralen Erkrankungen. Dtsch Gesundheitswes 31: 1804–1806
12. Payk T (1978) Therapie mit Psychopharmaka. Ther Ggw 117: 1328–1338
13. Pichlmayr I, Lips U (1979) Pethidin-Effekte im Elektroenzephalogramm. Anaesthesist 28: 433–442
14. Pichlmayr I, Lips U (1980) Promethazin-Effekte im Elektroenzephalogramm. Anaesthesist 29: 18–23
15. Pichlmayr I, Lips U (1980) Atropin-Effekte im Elektroenzephalogramm. Anaesthesist 29: 249–253

16. Pichlmayr I, Lips U (1980) Pethidin-Promethazin-Kombinationseffekte im Elektroenzephalogramm. Anaesthesist 29:254–260
17. Pichlmayr I, Lips U (1980) EEG-Effekte der Prämedikation mit Thalamonal®. Anaesthesist 29:360–365
18. Pichlmayr I, Lips U (1980) Diazepam-Effekte im Elektroenzephalogramm. Anaesthesist 29:317–327
19. Pichlmayr I, Lips U (1981) EEG-Effekte der Medikation mit Triflupromazin (Psyquil®). EEG EMG 12:105–107
20. Porcher H (1978) Le cerveau agé et ses médicaments: proceedings of a symposium. Selecta 29:2572–2580
21. Rovenstine EA, Battermann RJ (1943) Utility of demerol as substitute for opiates in preanesthetic medication. Anesthesiology 4:126–134
22. Schlungbaum H (1939) Schmerzbekämpfung mit Dolantin, einem synthetisch hergestellten Spasmolytikum und Analytikum. Med Klin 35:1259–1260
23. Schneiderlinn J (1903) Die Scopolamin-(Hyoszin-)Morphium-Narkose. Muench Med Wochenschr 1:371–372
24. Squibb Scientific information pamphlet on Psyquil (summary of references on triflupromazin)
25. Squires R, Braestrup C (1977) Benzodiazepine receptors in rat brain. Nature 266:732–734
26. Synder SH, Yamura HJ (1977) Opiate receptors and international opiates. Sci Am 236/3:44–56

II. Stages of Anesthesia

Narcotic, analgetic, and neuroleptic substances used in anesthesiology for general anesthesia cause a continuous reversible suppression of all or individual structures of the CNS (holencephalic and telencephalic narcotics). The changes in the level of consciousness and in the vegetative functions during the induction, maintenance, and termination of anesthesia are at all times related to the actual depth of anesthesia. The various stages of anesthesia can thus be clinically classified, as was first described by GUEDEL in 1920 [5]. In the course of anesthesia with holencephalic drugs, these stages are basically passed through in a similar sequence – analgesia with complete alertness, loss of consciousness, coma due to narcotic overdose, and intoxication. The time leading up to the individual stages and their duration depends on the type of narcotic agent and its specific appearance and disappearance in the CNS. In modern anesthetic techniques with combinations of several anesthetic-narcotic relaxing substances, the individual stages of anesthesia are frequently varied in their clinical manifestations and thus can be less easily distinguished than with the monoanesthetic formerly used.

Cerebral function is reflected in the EEG. Based on this fact, GIBBS et al. found in 1937 [4] that all anesthetics cause comparable EEG changes (today this is true for most anesthetics [9]). Thus, the depth of anesthesia can be determined by the EEG. The distinction of individual stages by EEG changes is much more exact than by clinically detectable changes. Based on the EEG classification of stages of sleep by LOOMIS et al. [10] and GUEDEL's clinical stages of anesthesia [5], SCHNEIDER and THOMALSKE [15], and similarly GIBBS and GIBBS (1951), introduced an EEG classification of anesthesia, which subsequently was made more precise by MARTIN et al. [11], KUGLER (1981), KUBICKI [6, 8], and KUBICKI et al. [7] (Tables 1, 2).

The individual stages of anesthesia are defined by EEG characteristics, determined by the type and degree of the cerebral depression (Tables 1, 2).

In the course of anesthesia, one defined stage glides smoothly into the next. Accordingly, the corresponding EEG pattern does not show changes abrupt. Sometimes, the frequency patterns, which are typical for different anesthetic stages, are found simultaneously. The various levels of consciousness during anesthesia and the corresponding EEG characteristics are shown in Table 2. The dominant frequency in the alert patient (mainly alpha activity, see Chap. AIV) during induction is first lowered then suppressed completely. Simultaneously, or immediately afterward, cerebral functions are activated. This is characterized by the appearance of fast waves and is clearly manifest by clinical signs of motor activity and psychic restlessness, or even excitation. Suppression, characterized by isolated slow waves, may also occur.

During the stage of somnolence, the slow frequencies become more marked; they form the pattern of so-called surgical anesthesia, when the anesthesia is further deepened. The stage of anesthetic coma is characterized by further deceleration of the frequency. With continuous medication of the anesthetic agent, the suppression – and suppression also of the deeper cerebral functions – is evident in electrocerebral silence (ECS), initially interrupted by bursts ("outbreaks of frequency"). These bursts become less frequent in the stage of complete anesthetic coma and eventually cease. The level trace in the EEG demonstrates the complete breakdown in brain function. This corresponds to the failure of all vegetative functions. At the same time, the tolerance limit for the anesthetic has been reached or exceeded. Even in the stage of complete anesthetic coma, a reversibility of the EEG changes and of the clinical condition is basically possible, however, this is dependent on the patient's general condition, the rapidity of the cerebral recovery from the substance, and the duration of the complete failure of brain function, with its negative effects on circulation and respiration. During recovery from anesthesia, the stages and the clinical and EEG signs appear in the reverse sequence to that seen during the application of anesthesia. During recovery – especially after the application of barbiturates – light stages of narcosis and fluctuations in the physiological stages of sleep are observed both clinically and encephalographically (KUGLER 1981). During the postnarcotic phase, an increase in the beta frequencies and/or a decrease in the amplitude of the initially dominant frequency can be demonstrated, even in the clinically alert patient (KUGLER 1981 [13, 14]). During anesthesia, the individual variations in the initial EEG disappear; in the deep stages of anesthesia, any local differences in the cortical activity become less noticeable [11].

The ability to determine exactly the anesthetic depth by the encephalographically recorded course of cerebral function described above are basic requirements for an EEG monitoring of anesthesia. This monitoring should be simplified, both from a theoretical and practical point of view, by decreasing the number of leads and measured values. The reduction of the number of recording leads from 18 to 2 still permits sufficient general changes in cerebral function to be detected, which is essential during anesthetic monitoring. The most advanced degree of reduction in the number of EEG leads has been achieved with the so-called integrated EEG – cerebral function monitor (CFM) [12]. Approximate values are derived for the frequency and the amplitude of a recorded and filtered one-lead EEG, which is traced as the integration graph of the total cerebral activity. SCHWILDEN and STOECKEL [16], on the basis of their EEG studies, regard

Table 1. Classification of stages of sleep, anesthesia, and coma according to clinical and electroencephalographic features

State (manner of assessment)	Initial situation	Classification of stages							Authors
Sleep (EEG)		A	B	C	D	E		–	LOOMIS et al. (1938)
Anesthesia (clinical)	Awake	I	II	III 1 | 2	3	4		IV	GUEDEL (1920)
Anesthesia (EEG)		I Induction		L Light anesthesia	M Moderate anesthesia	T Deep anesthesia		K Comalike anesthesia =TL (Tolerance limit)	SCHNEIDER and THOMALSKE (1956)
		Very light sleep		Light sleep	moderate sleep	Very deep sleep		–	SADOVE et al. (1967), GIBBS and GIBBS (1951)
	0	I		II	III	IV	V	VI	MARTIN et al. (1959)
	A_0	A_1A_2	$B_0B_1B_2$	$C_0C_1C_2$	$D_0D_1D_2$	$E_0E_1E_2$	F	–	KUGLER (1981)
	Normal	Stage of activation	Stage of falling asleep		Surgical stage of sleep and	Stage of narcotic deep sleep		Stage of complete narcotic deep sleep	KUBICKI (1968)
Intoxication (EEG)	Normal	Recovery phase		Somnolence	Coma (spontaneous reversibility)			Coma with collapse of vegetative function reversibility not impossible	KUBICKI et al. (1970)

Table 2. Changes and characteristics in the EEG according to the effect of the individual stages of anesthesia on CNS

Stages of anesthesia as determined clinically (GUEDEL [5])	Clinical changes: Respiration / Eye movement / Pupil size / Vomiting reflex / Eyelid reflex / Conjunctival reflex / Corneal reflex / Secretory reflex / Light reflex (GUEDEL [5])	EEG characteristic findings (KUBICKI [6])	
Initial state = awake	∅	Alpha (8–12 Hz)	 Normal
I	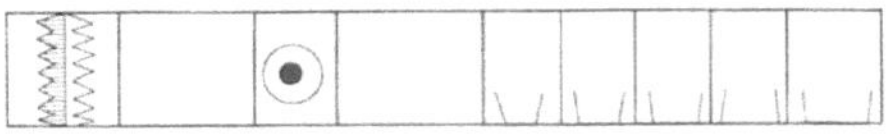	Alpha-Reduktion	 Stage of activation
II	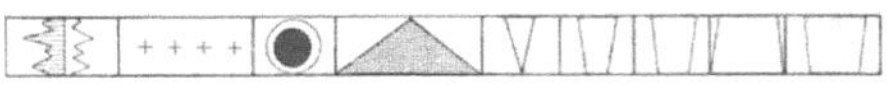	Beta (13–30 Hz)	 Stage of falling asleep
III 1	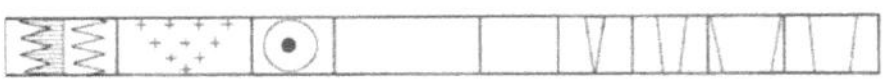	Delta-Theta (0,5–8 Hz) ↑ beta (13–30 Hz)	
2	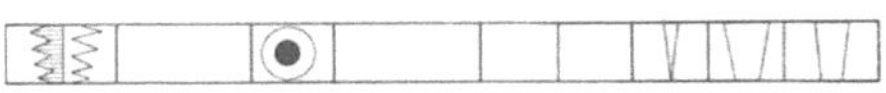	Delta-Theta (0,5–8 Hz)	 Surgical stage of sleep
3	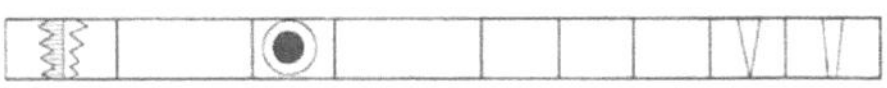	Delta (0,5–3 Hz)	
4	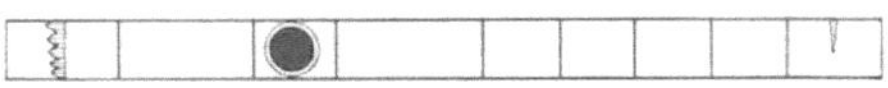	Delta (0,5–2 Hz) Burst-suppression	Stage of narcotic deep sleep 
IV	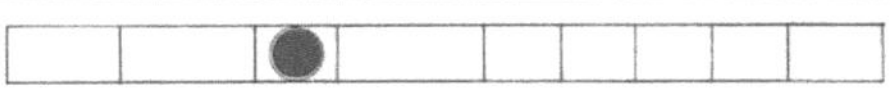	Suppression	Stage of complete narcotic deep sleep

aracteristics and spectrum of changes in the EEG ugler 1981)		Stages of anesthesia determined by the EEG findings (Kugler 1981)	CNS depression (review of literature in Nemes et al. 1979)	Classification of our own patients
riants		A_0	None	0
attened alpha	Alpha diffusion alpha low, sparse slow	A_1 A_2	Light cortical depression, euphoria, loss of powers of discrimination	1
ombined fast d underlying ow waves	Theta low delta low, theta moderately high	B_0 B_1 B_2	Cortical depression, unconsciousness, dominant subcortical hypersensitivity	2
ow, slow activity perimposed by w, fast waves	Theta high + delta 30% of the time, theta high 50% of the time, theta slow almost constant	C_0 C_1 C_2	Subcortical light depression	
rom moderate, gh to high, slow tivity	Delta up to 30% of the time, wide K complexes delta up to 50% of the time theta activity delta up to 80% of the time	D_0 D_1 D_2	Intense: mesencephalon dominates, onset of depression of medulla spinalis	3
igh, slow tivity	Delta constant delta extremely high delta extremely slow and high	E_0 E_1	Onset of mesencephalon depression	
mooth periodic ow bursts	Smooth, periodic slow bursts	E_2	Onset of pontine depression	4
mooth	Smooth	F	Depression of the medulla oblongata, collapse of vegetative function	5

the median of the power spectrum or the biparametrical graph of the median and mean amplitude as suitable EEG indicators of the anesthetic depth. Automatic management of anesthesia by means of analog frequency filter methods was introduced by BICKFORD in 1950 and has been used successfully in barbiturate anesthesia [1–3]. At present, both the described reduction in EEG information and the automatization of anesthetic control do not yet meet the demands of particular problems during anesthesia, and they require further improvement.

References

A. Textbooks and Summaries

Gibbs FA, Gibbs EL (1951) Atlas of Electroencephalography, vols. I–III. Addison-Wessly, Reading MA

Kugler J (1981) Elektroenzephalographie in Klinik und Praxis, eine Einführung, 3rd edn. Thieme Stuttgart

Nemes C, Niemer M, Noack G (1979) Datenbuch der Anästhesiologie. Fischer, Stuttgart

Sadove MS, Becka D, Gibbs FA (1967) Electroencephalography for anesthesiologist and surgeons. Pitmann, London

B. Papers

1. Bickford RG (1950) Automatic electroencephalographic control of general anesthesia. Electroencephalogr Clin Neurophysiol 2:93–96
2. Bickford RG (1951) Use of frequency discrimination in the automatic electroencephalographic control of anesthesia (Servo anesthesia). Electroencephalogr Clin Neurophysiol 3:83–86
3. Courtin RF, Bickford RG, Faulconer A (1950) The classification and significance of electroencephalographic patterns produced by nitrous-oxide-ether anesthesia during surgical operations. Proc Staff Meet Mayo Clin 25:197–206
4. Gibbs FA, Gibbs EL, Lennox WG (1937) Effect on electroencephalogram of certain drugs which influence nervous activity. Arch Intern Med 60:154–166
5. Guedel AE (1920) Signs of inhalation anesthesia. A fundamental guide. In: Guedel AE (ed) Inhalation anesthesia. Macmillan, New York, pp 10–52
6. Kubicki S (1968) Elektroenzephalographische Aspekte der Narkose. Berl Med 19:4–12
7. Kubicki S, Rieger N, Busse G, Barckow D (1970) Elektroenzephalographische Befunde bei schweren Schlafmittelvergiftungen. EEG EMG 1:80–93
8. Kubicki S (1971) EEG-Veränderungen durch Neuroleptanalgesie. In: Henschel WF (ed) Neue klinische Aspekte der Neuroleptanalgesie. Schattauer, Stuttgart, p 37–42
9. Kubicki S, Herrmann WM, Fischte K, Freund G (1979) Reflection of the topics: EEG frequency bands and regulation of vigilance. Pharmakopsychiatr Neuropsychopharmakol 12:237–245
10. Loomis AL, Harvey EN, Hobart CA (1938) Distribution of disturbance patterns in the human electroencephalogram with special reference to sleep. J Neurophysiol 1:413–430
11. Martin JT, Faulconer A, Bickford RG (1959) Electroencephalography in anesthesiology. Anesthesiology 20:359–376
12. Maynard D, Prior PF, Scott DF (1969) Device for continuous monitoring of cerebral activity in resuscitated patients. Br Med J 4:545–546
13. Pichlmayr I, Gubernatis G, Luba A (1977) EEG-Verhalten in der Aufwachphase nach barbituratinduzierter Neurolept-Analgesie. 15th Gemeinsame Tagung der Deutschen – Österreichischen und Schweizerischen Gesellschaften für Anästhesiologie und Reanimation, 13–17 Oct 1977. Geneva

14. Pichlmayr I, Gubernatis G, Luba A (1977) EEG-Verhalten in der Aufwachphase nach barbituratinduzierter Halothannarkose. 15th Gemeinsame Tagung der Deutschen – Österreichischen und Schweizerischen Gesellschaften für Anästhesiologie und Reanimation, 13–17 Oct 1977. Geneva
15. Schneider J, Thomalske G (1956) Betrachtungen über den Narkosemechanismus unter besonderer Berücksichtigung des Hirnstammes. Zentralbl Neurochir 16: 185–202
16. Schwilden H, Stoeckel H (1980) Untersuchungen über verschiedene EEG-Parameter als Indikatoren des Narkosezustandes. Anaesth Intensivther Notfallmed 15: 279–286

III. Inhalation Anesthetics

CONTENTS

In modern combination anesthesia, inhalation methods are either employed for the main anesthesia or in low-dose intermittently used adjuncts of anesthesia. Pure inhalation anesthesia is only applied in cases with special indications, most frequently in pediatric anesthesiology. The inhalation anesthetics used today include – nitrous oxide (laughing gas), halothane, enflurane, methoxyflurane, isoflurane, and fluroxene. Ethyl chloride, chloroform, diethyl ether, cyclopropane, and trichloroethylene are hardly used today due to their disadvantages either in clinical effects or application. Of the various substances available, each hospital normally uses one or two inhalation anesthetics according to its particular requirements in addition to nitrous oxide, which is used as a basic analgetic. In this chapter, therefore, our own findings are described only for halothane and enflurane.

The pharmacokinetics of the various inhalation anesthetics is determined by the applied concentration, the alveolar ventilation, and the solubility in the blood and tissues. The lower the solubility, the more rapidly a high blood and tissue level will be achieved. The depth of anesthesia depends on the level of the anesthetic in the brain. As a measure for the narcotic strength of a substance, and as an approximate value for its dosage, so-called MAC values were introduced [12] and determined for inhalation anesthetics (MAC = minimal alveolar concentration of the substance sup-

Table 1. Physical data of the most commonly used inhalation anesthetics (after GOODMAN and GILMAN 1980)

Inhalation anesthetic	Steam pressure (mm Hg at 20 °C)	Blood/gas distribution coefficient at 37 °C	Lipid/gas distribution coefficient at 37 °C	MAC[a] (Vol. %)
Nitrous oxide	Exists only in gaseous phase at room temperature	0.47	1.4	105[b]
Halothane	243	2.3	224	0.75
Enflurane	175	1.9	98	1.68
Methoxyflurane	22.5	12.0	970	0.16
Isoflurane	250	1.4	99	1.40

[a] MAC, minimal alveolar concentration
[b] An MAC value above 100 vol.% means that MAC can only be achieved under hyperbaric conditions

pressing pain perception in 50% of patients at a gas pressure of 1 bar) (Table 1). Even today, the clinical effects of inhalation anesthetics are determined by comparison with ether, which when slowly applied allows distinct anesthetic stages to be recognized (BENZER et al. 1982; DRIPPS et al. 1977; GOODMAN and GILMAN 1980; KUSCHINSKY and LÜLLMANN 1981; LEE and ATKINSON 1978; NEMES et al. 1979). The changes visible in the EEG during anesthesia with the various substances show both similar features (EEG-defined stages) and specific differences. Inhalation anesthetics decrease cerebral metabolism and, dependent on dose – at least in the application phase – increase cerebral perfusion and cerebrospinal fluid (CSF) pressure by dilatation of the cerebral vessels. The autoregulation of the cerebral perfusion remains essentially intact [5, 6, 17, 18, 20–25, 31–34, 37].

Nitrous Oxide (Laughing Gas)

Pharmacological Mode of Action and Clinical Application. Nitrous oxide is an anesthetic gas, which in its clinical application with an oxygen content of at least 20% has weak anesthetic but pronounced analgetic effects. After physical solution of the gas in the blood, the central nervous effects are not primarily due to metabolic processes, but are probably caused by a mechanical displacement of oxygen in the ganglion cells. Chemical reactions of nitrous oxide combined with sufficient oxygen and with a limited application time is thought to be almost harmless. It is used as a basic analgetic in almost all general anesthesia. Its application obviates the need for further narcotics with their concomitant side effects.

EEG Findings (Table 2)

Our Findings. Owing to the combined application of nitrous oxide with other anesthetics, changes in the EEG specifically due to nitrous oxide cannot be established. The inhalation of nitrous oxide alone cannot be justified.

Review of Other Findings. The inhalation of nitrous oxide/oxygen in the ratio of 4:1 leads, in accordance with the limited narcotic effectivity, to light anesthesia and to the respective EEG stages down to C_1. The EEG grades 1–3 were detected by some authors [2, 29] BRECHNER et al. 1962; SADOVE et al. 1967); grade 1 was not found by the majority of authors. With limited oxygen content and proportionally increased nitrous oxide, anesthesia may be increased to stage III_2 (D_1) [13]. With curarization and nitrous oxide/oxygen (ratio 2:1) administration over a longer period (90 min), high-amplitude spindle-shaped alpha activity is observed (BRECHNER et al. 1962; SADOVE et al. 1967).

Halothane (Fluothane)

Pharmacological Mode of Action and Clinical Application. Halothane, first produced in 1951, is highly effective and easily regulated; today it is the major and, apart from nitrous oxide, the most commonly used inhalation anesthetic. The analgetic component of the substance has little effect; narcotic actions and side effects are dose-dependent and mainly involve a marked decrease in blood pressure – due to peripheral vasodilatation and cardiodepression. The MAC value for halothane is 0.75 vol%. Cerebral metabolism is decreased, cerebral perfusion – at least over the first 30–60 min – increased, with almost constant peripheral blood pressure readings. The CSF pressure is raised accordingly. Cerebral autoregulation is generally preserved, but is slowed down according to the extent of cerebral vasodilatation. Halothane anesthesia produces a cerebral "luxury perfusion" under stable peripheral circulation [21, 34]. A metabolic rate of 10–25% and degradation into very reactive metabolites may be the causes of immunological hepatic damage (BENZER et al. 1982).

EEG Findings (Table 2; Fig. 1)

Our Findings. (50 patients 20–90 years of age [35]). The initial EEG (dominant frequency (DF), 8–12 Hz, 20–60 μV in 80% of the patients; normal variance without pathological significance in 20% of the patients) is modified by inhalation of 1 vol.% of halothane and nitrous oxide/oxygen (ratio 3:1) according to the stage of anesthesia. Within 20 min, 38 of 50 patients attain anesthetic stage III_{2-3} (D_{1-2}); 12 patients attain stage $II–III_1$ ($B_2–C_1$). In the first few minutes, the dominant initial frequency is gradually suppressed. There then follow fast frequencies, which are subject to a frequency deceleration from the fast to the slow beta range with further deepening of anesthesia; with the transition to stage III (C), they gradually become completely reduced. At amplitude greater than 20 μV, the fast frequencies are associated with clinical signs of excitation. Strong premedication, a poor general state of health, decreased blood pressure, or initial high levels of halothane inhibit or diminish the occurrence of fast waves. In stage III (CD) of anesthesia, slow waves increase in number and amplitude. This occasionally occurs in stage II. The stage III_2 level of anesthesia was not exceeded among our own patients with further frequency deceleration and increase in amplitude.

Review of Other Findings. There is good agreement between our own findings and those of other authors on the EEG of halothane anesthesia [7, 16, 26] (BRECHNER et al. 1962; SADOVE et al. 1967). In stage III_3 (E), very slow

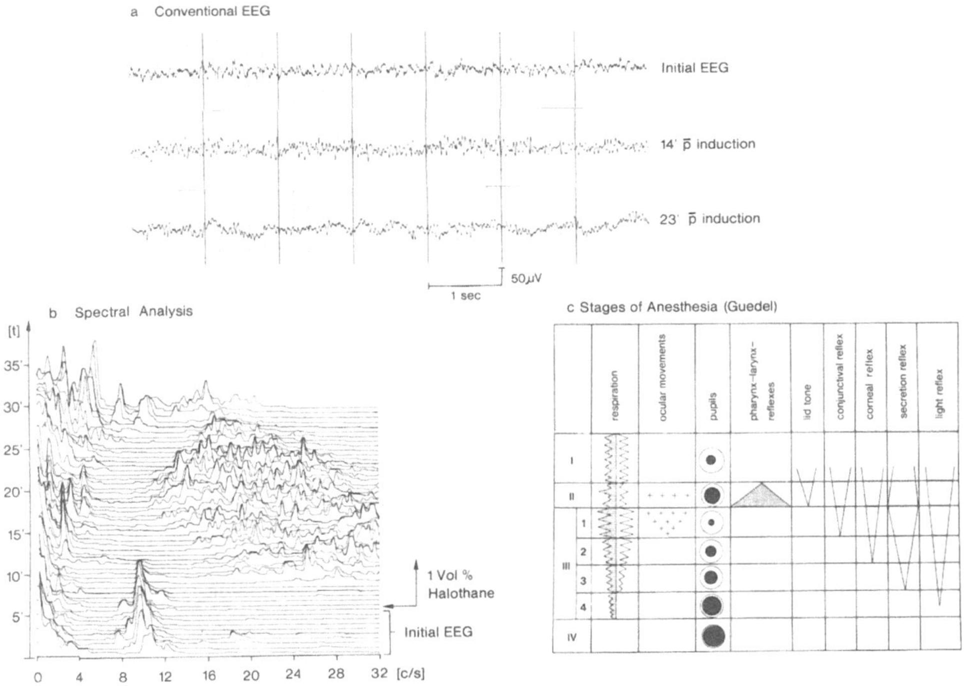

Fig. 1a–c. Typical EEG pattern in two different graphs with halothane induction (halothane 1 vol.%; $N_2O/O_2 = 3:1$). **a** Conventional EEG. Initial alpha EEG, 25–30 μV; dominant frequency 9 Hz; 14 min after halothane – beta EEG approx. 15–20 μV; dominant frequency approx. 25 Hz; 23 min after halothane-delta EEG approx. 75 μV, dominant frequency 2–3 Hz, some interspersed beta activity. **b** Spectral picture of the EEG. Halothane-specific passing of the Guedel stages I–III/2. Frequency pattern corresponding with conventional EEG. **c** Stages of anesthesia according to Guedel.
Recording conditions: **a** lead C_3–P_3; TC, 0.3 s; filter, 70 Hz; calibration 50 μV = 14 mm; paper speed, 30 mm/s. **b** lead C_Z–A_1; FFT, 30-s periods

Table 2. Summary of EEG findings from the literature for inhalation anesthesia

Anesthetic agent, Authors	Parameters observed	EEG characteristics with defined stages of anesthesia,		
			I	II
		A_0	A_1 A_2	B_0 B_1 B_2
		0	1	2
General	Frequency	8–13		↑ 20–30
MARTIN et al. [26]	Electric power	5–50		50
BRECHNER et al. (1962)	Others	Variants without pathological value		
Nitrous oxide	Frequency		=	18–24
BEECHER and MCDONOUGH [2]	Electric power		↓	< 50
FAULCONER et al. [13]				
PEARCY et al. [29]	Others			
SADOVE et al. (1967)				
Halothane (+ N_2O/O_2)	Frequency			15–25
GAIN and PALETZ [15]	Electric power			10–25
	Others			
BRECHNER et al. (1962)	Frequency			15–20
	Electric power			10–25
	Others			
SADOVE et al. (1967)	Frequency			15–25
	Electrical power			Low
	Others			
STOCKARD et al. [38]	Frequency			24
	Electrical power			40–60
	Others			
PICHLMAYR and LIPS [35]	Frequency	8–12		15–25
	Electrical power	20–60	↓ to 0	10–50
	Others	irregular with pathological values		
Enflurane (+ N_2O/O_2)	Frequency			35
BART et al. [1]	Electric power			
	Others			
CLARK et al. [7]	Frequency			14–18
	Others			

as determined electroencephalographically and clinically[a]

	III$_1$	III$_2$	III$_3$	IV	According to Guedel (1920)
	C_0 C_1 C_2	D_0 D_1 D_2	E_0 E_1 E_2 F		According to Kugler (1966)
		3	4	5	Our classification
	1–5 (Regular) 50–300	1–3 (Irregular) + 3–6 100–300 50–100	1–2 ↓		Hz μV
			Burst-suppression (= periodic interruption of cortical activity, of increasing duration)	Suppression = depression = complete loss of cortical activity	
	2–4 (–7) ↑	2–4 ↓ ↑			Hz μV
	3–6 → 4 50 50–100 Superimposed fast – frequency of slow amplitude	2–3 (Irregular) 100–300	1 + 6–8 100 – 200 25 + 50 Burst-suppression	Isoelectric EEG	Hz μV
	4 50–100	2–3 100–300	1 and 6 100–300 and 25– 40		Hz μV
	4–6 Medium	3+4–5 High	2–4 (Irregular) Low Further ↓ Further ↑		Hz μV Hz μV
			periodic sharp waves		
	3–5 50–100 Restriction of upper frequency limit	2–3 50–200			Hz μV
	3–10 80	8–10 80 Spikes 150–200 μV	Spikes and burst-suppression		Hz μV μV
	Spike-dome complexes, later alternating with electric silence				

[a] ↑ increase, ↓ decrease, ↑↓ increase and decrease

Table 2 (continued)

Anesthetic agent, Authors	Parameters observed	EEG characteristics with defined stages of anesthesia,		
			I	II
		A_0	A_1 A_2	B_0 B_1 B_2
		0	1	2
Neigh et al. [27]	Frequency Electric power E concentration Others			15–20 25 1
Bosken et al. [3]	Frequency Electric power E concentration Others		=	20 1
Gies et al. [16]	Frequency Electric power Others	Similar frequency pattern to that with halothane. Early increase in amplitude and sharp waves with slow postfluctuation (2.5 Vol. %, no spike/wave complex)		
Doenicke and Kugler [11]		Generally similar to classic EEG stages of inhalation anesthetics; frequent occurrence of sharp waves		
Our findings	Frequency Electric power Enflurane concentration Others	8–12 20–60 Irregular forms with pathological value	= Limitation of upper frequency limit (all stages)	13–32 25–30 =
Methoxyflurane (+ N_2O/O_2) Sadove et al. (1967) Bart et al. [1] Yurewich [39]	Frequency Electric power Others			18–25 ↓
Isoflurane (+ N_2O/O_2) Eger et al. [12]	Frequency Electric power Others			↑ ↓
Fluroxene (+ N_2O/O_2) Brechner and Dornette [4] Sadove et al. (1967) Bart et al. [1]	Frequency Electric power Others	8–13 20–50	Similar changes to those with ether anesthesia; frequency deceleration with deepening of anesthesia	20–30 59
Cyclopropane (+ N_2O/O_2) Possati et al. [36] Brechner et al. (1962) Sadove et al. (1967)	Frequency Electric power Others	8–13 50 Similar changes to those with ether; small amplitude in deeper stages of anesthesia; EEG findings correlate with blood levels		↑ 20–30 ↓ 20

as determined electroencephalographically and clinically[a]

III_1	III_2	III_3	IV	According to Guedel (1920)
C_0 C_1 C_2	D_0 D_1 D_2	E_0 E_1 E_2 F		According to Kugler (1966)
3		4	5	Our classification
2–4 (7)	2–4 (7)			Hz
50–100	50–100			μV
2	2,5–3	3.5		Vol. %
Superimposed fast waves	Spikes 150 μV	Spikes (150–300 μV) and burst-suppression		
4–7	4–5			Hz
				μV
2	2.5	Burst-suppression		Vol. %
	Polyspikes and waves			
10				Hz
High amplitude				μV
	Polyspikes and waves 500 μV	Burst-suppression		
2–7	2–6			Hz
50–100	100			μV
	=			Vol. %
Occasionally superimposed fast activity				
9–15	4– 7	1–3		Hz
	12–15			μV
40–60	> 100	80–120		μV
				Hz
14–17				μV
120–180				
		Burst-suppression		Hz
3–4	3	1–2		μV
50–100	150	150–300		
3–8	2–8	1–2		Hz
↑ 100–200	= ↓ 50–20			μV
		Burst-suppression		

Table 2 (continued)

Anesthetic agent, Authors	Parameters observed	EEG characteristics with defined stages of anesthesia,		
			I	II
		A_0	A_1 A_2	B_0 B_1 B_2
		0	1	2
Trichloroethylene (+ N_2O/O_2)	Frequency			↑
BEECHER and MCDONOUGH [2]	Electric power			↓
COURTIN [9]	Others	EEG pattern similar to that with other volatile substances		
BRECHNER et al. (1962)				
Ethyl chloride		Decrease in frequency and increase in amplitude with deepening anesthesia; no longer used as general anesthetic		
UJÜC [39]				
Chloroform	Frequency	Beta, occasionally delta		20–30
PEARCY et al. [30]	Electric power	↓		↑
Diethyl ether (+ N_2O/O_2)	Frequency		Alpha	20–30
COURTIN et al. [8]	Electric energy		↓ up to 0	Low
	Others			
FAULCONER [14]		Ether blood levels show a linear response parallel to EEG changes		
BRECHNER et al. (1962)	Frequency	8–13	5	20–30
	Electric energy	25–50		30
	Others	Superimposed fast waves of low amplitude		
SADOVE et al. (1967)	Frequency	10		15–25
	Electric energy			Low
	Others			

high waves predominate. These may change into burst-suppression phases and, according to GAIN and PALETZ, result in the cessation of cortical activity [15].

Enflurane (Ethrane)

Pharmacological Mode of Action and Clinical Application. Enflurane was introduced as an inhalation anesthetic for clinical use in 1973. Its physical

as determined electroencephalographically and clinically[a]

III_1	III_2	III_3	IV	According to Guedel (1920)
C_0 C_1 C_2	D_0 D_1 D_2	E_0 E_1 E_2 F		According to Kugler (1966)
	3	4	5	Our classification
3–5 (Rhytmic)	2–3 (Arrhythmic)			Hz
100–200	>200			μV
–4 Irregular	2–7 Irregular	2–4		Hz
↓	=	↓		μV
–8 (Rhythmic)	2–5 (Arrhythmic)			Hz
–300	>200			μV
	Superimposed fast waves	Burst-suppression Mild 3 s Moderate 3–10 s Severe 10 s	Isoelectric EEG	Duration of iso-electric phases
6	1–3+ 4–8	2–3 1 (Irregular)		Hz
200–300	150–200 +50–100	150 100		μV
		Burst-suppression		
	1–3			Hz
(Rhythmic) High	(Arrhytmic) High			μV
		Slight moderate marked complete suppression		

characteristic – fast absorption and breakdown in the tissues – allow easy regulation as seen in Table 1. The MAC value of enflurane is usually given as 1.68 vol.%. Anesthesia with enflurane causes a dose-dependent reversible cardiogenic depresssion and, through vasodilatation, a decrease in peripheral vascular resistance. The subsequent dose-dependent decrease in blood pressure corresponds with that of equipotential halothane application. In contrast to halothane, enflurane is said to cause no bradycardia and no sensitization of the heart. Of the applied enflurane, 2–5% is metabolized in the liver. Fluorides, as metabolic products, have been thought to have a negative effect on renal disease. Enflurane restricts cerebral metabolism; cerebral perfusion is increased, with the perfusion pressure remaining constant; CSF pressure increases accordingly.

EEG Findings (Table 2; Fig. 2)

Our Findings (16 patients; eight patients under 50 years of age, eight patients over 70 years). At the onset of anesthesia, the effects of enflurane inhalation correspond with the anesthetic stages that have already been clinically and encephalographically described under halothane. The background activity is at first suppressed, then, within a few minutes, this is replaced by beta activity. Beta frequencies are generally faster with enflurane than with halothane; they become slower when anesthesia is increased. In the stage of surgical anesthesia, III_1–III_2 (CD), high-voltage delta/theta activity occurs, which at the beginning of stage III_1 (C_0–C_1) is superimposed by additional fast frequencies. We did not attempt to subject our patients to deeper stages of anesthesia. Considerable dose-dependent decreases in blood pressure require, as well as volume replacement, the occasional application of sympathicomimetics in order to maintain constant cerebral perfusion. In contrast to our findings with halothane, in stage II (B) the electric power in the beta spectrum cannot be related to clinically manifest excitation. Changes in the Ethrane level in the anesthesia lead to visible changes in anesthetic depth in the EEG more rapidly than with halothane. This confirms the good ability to regulate the substance.

Review of Other Findings. The literature confirms that the EEG changes with enflurane are basically comparable with those of other inhalation anesthetics [1, 3, 11, 16, 27]. The frequent occurrence of sharp waves, an early visible increase in amplitude, and a quick reversibility of the changes are regarded as special features of enflurane anesthesia. Very high concentrations of enflurane (2.5 vol.%), which would be too dangerous for clinical use due to the cardiac and circulatory effects, would in stage III_3 lead to paroxysmal polyspike-wave complexes of high amplitude, which could be enhanced by hyperventilation. These together with occasional tonoclonic motoractivity could give rise to an epileptogenic effect of the substance. The incidence of epileptic seizures with enflurane, however, does not exceed the expected rate of provocative activation in the average population [28]. Although myoclonias are regarded as a disinhibition of subcortical structures and sharp waves are thought to be special processing of afferent impulses [19], the application of enflurane in epileptics should nonetheless be undertaken with great care [10]. As with other substances, burst-suppression phases with enflurane are a manifestation of overdose and must be interpreted as a sign of limited cerebral tolerance toward the substance.

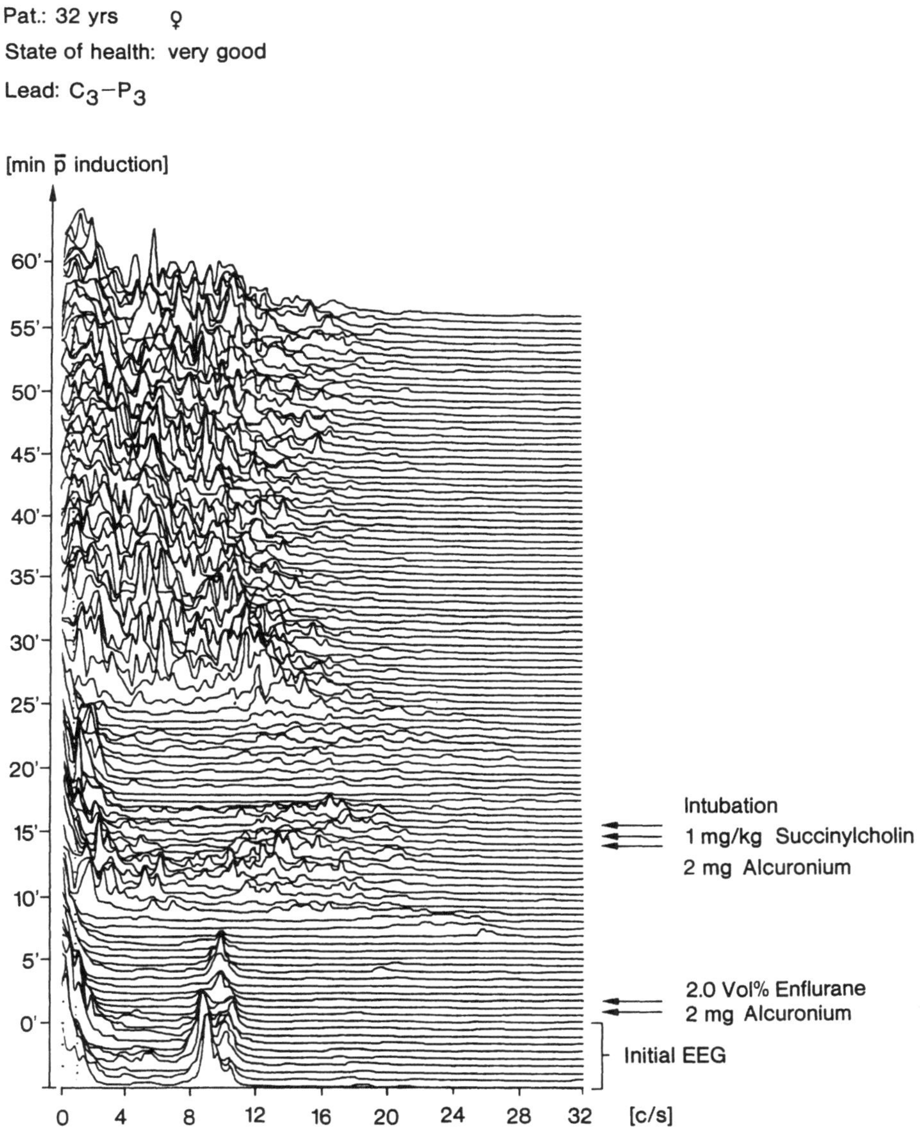

Fig. 2. EEG course of enflurane induction continued into light anesthesia. The well-marked alpha activity disappears slowly after enflurane application. Activities occur mainly in the beta range and in the delta/theta ranges. Beta fractions disappear after 7–10 min; delta activity remains. After 25 min, reduction of enflurane. As a manifestation of the light anesthesia, there is a high-voltage irregular EEG with an upper border frequency of 18 Hz (anesthetic stages according to Guedel, transition II–III).
Recording conditions: lead C_3–P_3; TC, 0.3 s; filter: 70 Hz; calibration, 50 μV = 7 mm; FFT in 30-s periods

Methoxyflurane (Pentrane)

Pharmacological Mode of Action and Clinical Application. Methoxyflurane has been used clinically since 1960. Despite its high potency with a MAC value of 0.16 vol.%, and notwithstanding its high blood solubility, it is absorbed and broken down slowly. Methoxyflurane has dose-dependently cardiodepressive effects. The peripheral vascular resistance is not changed. Fluorides may interfere with renal function through biotransformation. Like other inhalation anesthetics, methoxyflurane reduces cerebral metabolism; simultaneous dilatation of the brain vessels leads to an increase in cerebral perfusion and intracranial pressure.

EEG Findings (Table 2)

Review of Findings in the Literature. As with other substances, beta frequencies of low amplitude are observed with methoxyflurane in stage II (B). In contrast to inhalation anesthetics discussed above, with an increase in anesthesia in stage III_1 (C_0–C_1) alpha frequencies occur once more, which then slow down through the theta range in stage III_2 (D) to the delta range (stage III_3–I) [40].

Isoflurane (Forane)

Pharmacological Mode of Action and Clinical Application. Isoflurane is similar to enflurane in its physical properties and clinical mode of action; it is an isomer of the enflurane molecule. Excitatory phenomena are rare. In contrast to other inhalation anesthetics, cardiodepressive effects are negligible. A marked dose-dependent decrease in blood pressure occurs due to peripheral vasodilatation. Possible negative effects on the liver, e.g., tumor induction await further clarification.

EEG Findings (Table 2)

Review of Findings in the Literature. Isoflurane causes the expected increase in frequency in stage II (B). A peculiarity of isoflurane is that low beta waves are observed in stage III_1 (C). With overdose, burst-suppression phases occur [12].

Fluroxene (Fluromar)

Pharmacological Mode of Action and Clinical Application. Fluroxene was developed in 1951. Rapid action and brief narcotic hangovers are due to its low solubility coefficient. Fluroxene is associated a high incidence of postoperative nausea and vomiting.

EEG Findings (Table 2)

Review of Findings in the Literature. With fluroxene changes in the EEG are seen similar to those with other inhalation anesthetics. First, there is a stage of increased frequency with a simultaneous decrease in amplitude. With further intensification of anesthesia, the frequency slows down to the delta range, with another increase in amplitude [1].

The following inhalation anesthetics are mainly of theoretical interest; they are no longer used either because of their possible hazard or their proven side effects. Only in exceptional cases are they used clinically.

Cyclopropane (Trimethylene)

Pharmacological Mode of Action and Clinical Application. As an inhalation anesthetic, cyclopropane was introduced in 1934. It is highly explosive and could no longer be used in the operating room when electric devices were introduced. Cyclopropane has a low distribution coefficient, high potency, and an wide therapeutic range. It depresses respiration more than other inhalation anesthetics. Cyclopropane sensitizes the heart against catecholamines and lowers cardiac contraction in a dose-dependent manner. The latter effect is masked by activation of central vasomotor and vasopressor nuclei. Both sympathetic and vagal tones are increased. Clinically, an increase in blood pressure and bradycardia occur. Pronounced excitations are seen when the patient falls asleep or awakens; postnarcotically, the incidence of nausea and vomiting is high.

EEG Findings (Table 2)

Review of Findings in the Literature. Again, with cyclopropane, the synchronization of frequency indicates the deepening of anesthesia after a phase of nonsynchronization. Burst-suppression phases, again, are of overdose [36].

Trichloroethylene (Trichlorethen)

Pharmacological Mode of Action and Clinical Application. Trichloroethylene was clinically introduced in 1934; it is hardly used nowadays. With a high bloodgas distribution coefficient, anesthetic induction and recovery are very slow. Trichloroethylene is only feasible as a component of combination anesthesia, since at deeper stages of anesthesia it leads to severe respiratory and cardiogenic depression. The use of the substance in recirculating respiratory systems is not possible because of its toxic effect in connection with heated CO_2-absorbent lime.

EEG Findings (Table 2)

Review of Findings in the Literature. EEG changes under trichloroethylene anesthesia have been recorded down to stage III_{2-3} (CD). They demonstrate the usual course of inhalation anesthesia [9].

Ethyl Chloride (Chlorethyl)

Pharmacological Mode of Action and Clinical Application. Ethyl chloride, which acts quickly, was used until about 1960 for induction and for producing analgesia during rapid surgery ("quick rausch"). Severe cardiotoxic side effects ruled out the possibility of deepening of anesthesia and its further use today. Only the local anesthetic effects of icing with ethyl chloride is still used.

EEG Findings (Table 2)

Review of Findings in the Literature. In stages II–III of anesthesia with ethyl chloride, the EEG shows a frequency decrease combined with increasing amplitude [38].

Chloroform

Pharmacological Mode of Action and Clinical Application. Chloroform is no longer used; it is an inhalation anesthetic with a small narcotic range and cardiohepato- and nephrotoxic properties.

EEG Findings (Table 2)

Review of Findings in the Literature. Effects of chloroform anesthesia on the EEG have been recorded down to stage III_3 (D); they correspond with the usual changes seen with other inhalation anesthesias [20].

Diethyl Ether (Ether purissimum, pro narcosi)

Pharmacological Mode of Action and Clinical Application. Ether is the oldest inhalation anesthetic. It has been known since the sixteenth century; its first famous clinical application was in 1846, by MORTON. The substance has a high blood solubility and acts slowly, thus, a wide therapeutic range results. The anesthesia is accompanied by good analgesia due to central depression, progressing from the brain to the medulla. Cardiac or circulatory impairments are small down to the level of surgical anesthesia with simultaneous sympathicomimetic effects. In deeper stages of anesthesia, which are characterized by muscular relaxation, the blood pressure falls due to paralysis of medullary circulatory centers. The high incidence of postoperative nausea and vomiting is characteristic of ether anesthesia. Despite its wide therapeutic range and positive analgetic narcotic properties, the explosiveness of the anesthetic mixture is responsible for its almost complete disappearance from clinical practice.

EEG Findings (Table 2)

Review of Findings in the Literature. The individual stages of anesthesia can be observed especially well due to the slow anesthetic action, which parallels the clinical symptoms and effects on the EEG. After the disappearance of the dominant initial frequency in stage I ($A_{1,2}$), beta activity of low amplitude is observed in stage II (B), which initially changes into a rhythmic delta/theta activity of high amplitude (III_1; C). In stage III_2 (D), arrhythmic slow and high waves predominate. The stage III_3 (E) is characterized by an increased activity of burst-suppression phases. With continuation of ether inhalation, cerebral activity is initially reversible at this stage, but then becomes irreversibly suppressed [8, 30] (BRECHNER et al. 1962; SADOVE et al. 1967). The ether levels measured in the blood show a pattern that corresponds with the EEG stages of anesthesia [14].

Summary of EEG Findings with Inhalation Anesthetics

Although the various inhalation anesthetics are more or less different due to their varying times of onset and the resulting clinical symptoms, they exhibit similar EEG changes, which determine the individual anesthetic stages, parallel to the clinical observations. The excitation stage II, which appears clinically only in a reduced form, or not at all with the new inhalation anesthetics, can be clearly determined from the EEG. The particular value of recording cerebral anesthetic depth is in avoiding overdose. It also enables the anesthesiologist to avoid a premature surgical cut under muscular relaxation and to maintain a particular anesthetic stage at a constant level.

References

A. Textbooks and Summaries

Benzer H, Frey R, Hügin W, Mayrhofer O (1982) Lehrbuch der Anästhesiologie, Intensivmedizin und Reanimation. Springer, Berlin Heidelberg New York

Brechner VL, Walter RD, Dillon JB (1962) Practical electroencephalography for the anesthesiologist. Thomas, Springfield

Dripps RD, Eckenhoff JE, Vandam LD (1977) Introduction to anesthesia. The principles of safe practice. Saunders, Philadelphia

Goodman LS, Gilman A (1980) The pharmacological basis of therapeutics. McMillan, New York

Kugler J (1966) Elektroenzephalographie in Klinik und Praxis. Thieme, Stuttgart

Kuschinsky O, Lüllmann H (1981) Kurzes Lehrbuch der Pharmakologie. Thieme, Stuttgart

Lee JA, Atkinson RS (1978) Synopsis der Anaesthesie. Fischer, Stuttgart

Nemes C, Niemer M, Noack G (1979) Datenbuch der Anästhesiologie. Fischer, Stuttgart

Sadove MS, Becka D, Gibbs FA (1967) Electroencephalography for anesthesiologists and surgeons. Pitham, London

B. Papers

1. Bart AJ, Homi J, Linde HW (1971) Changes in power spectra of electroencephalograms during anesthesia with fluroxene, methoxyflurane and ethrane. Anesth Analg Curr Res 50:53–63
2. Beecher HK, McDonough FK (1939) Cortical action potentials during anesthesia. J Neurophysiol 2:289–307
3. Bosken F, Hanquet M, Galletz JP (1974) Enflurane and EEG. Acta Anaesthesiol Belg 2:233–245
4. Brechner VL, Dornette WHL (1957) Electroencephalographic patterns during nitrous-oxide-trifluorethyl vinylether. Anesth Analg 18:321–327
5. Christensen MS, Hoedt-Rasmussen K, Lassen NA (1965) The cerebral blood flow during halothane anesthesia. Acta Neurol Scand [Suppl] 14:152–152
6. Christensen MS, Hoedt-Rasmussen K, Lassen NA (1967) Cerebral vasodilatation by halothane anesthesia in man and its potentiation by hypotension and hypercapnia. Br J Anaesth 39:927–934
7. Clark DL, Hosik EC, Rosner BS (1971) Neurophysiological effects of different anesthetics in unconscious man. J Appl Physiol 31, 6:884–891
8. Courtin RF, Bickford RG, Faulconer A (1950) The classification and significance of electroencephalographic patterns produced by nitrous oxide ether anesthesia during surgical operations. Proc Staff Meet Mayo Clin 25:197–206
9. Courtin RF (1955) Electroencephalographic and clinical observations with trichlorethylene and nitrous oxide. Anesthesia (Dallas) MJ 41:613–615
10. Degen R (1980) Diskussionsbemerkung auf dem Symposium über Anästhesie bei zerebralen Krampfanfällen, Bielefeld 1979. In: Opitz A, Degen R (eds) Anästhesie bei zerebralen Krampfanfällen und Intensivtherapie des Status epilepticus. Perimed, Erlangen
11. Doenicke A, Kugler J (1975) Wirkungen des Ethrane auf das zentrale Nervensystem. In: Kreuscher H (ed) Ethrane. Neue Ergebnisse in Forschung und Klinik. Schattauer, Stuttgart, pp 45–55
12. Eger EJ, Stevens WG, Cromwell TH (1971) The electroencephalogram in man anesthetized with forane. Anesthesiology 35(5):504–508
13. Faulconer A, Pender JW, Bickford RG (1949) Influence of partial pressure of nitrous oxide on depth of anesthesia and electroencephalogram in man. Anesthesiology 10:601–609
14. Faulconer A (1952) Correlation of concentrations of ether in arterial blood with electroencephalographic patterns occurring during ether anesthesia of human surgical patients. Anesthesiology 13:361–369
15. Gain EA, Paletz SG (1957) An attempt to correlate the clinical signs of fluothane anesthesia with the electroencephalographic levels. Can Anaesth Soc J 4:289–294
16. Gies B, Gerking P, Scholler KL (1974) Das EEG bei Probanden-Narkosen und kontinuierlicher EEG-Frequenzanalyse (EISA) während Operationen unter Ethrane. In: Lawin P, Beer R (eds) Ethrane. Anästhesiologie und Wiederbelebung. Springer, Berlin Heidelberg New York, p 263–271
17. Kekeski F, Gallyas F, Szantano J (1967) Die Wirkungen von verschiedenen Inhalationsnarkosetypen auf die Gehirndurchblutung. Acta Med Acad Sci Hung 24:153–161
18. Kreuscher H, Grote J (1969) Die Hirndurchblutung und cerebrale Sauerstoffaufnahme in Narkose. In: Betz E, Wüllenweber R (eds) Pharmakologie der Gehirndurchblutung. Ärztl Forsch Sonderbd, p 120–124

19. Kugler J (1980) Diskussionsbemerkung auf dem Symposium über Anästhesie bei zerebralen Krampfanfällen, Bielefeld 1979. In: Opitz A, Degen R (eds) Anästhesie bei zerebralen Krampfanfällen und Intensivtherapie des Status epilépticus. Perimed, Erlangen, p 162–163
20. Lassen NA, Hoedt-Rasmussen K, Christensen MS (1969) Halothane: A cerebral vasodilatator drug. In: Betz E, Wüllenweber R (eds) Pharmakologie der lokalen Gehirndurchblutung. Ärztl Forsch Sonderbd, p 111–115
21. McDowall DG (1967) The effects of clinical concentrations of halothane on the blood flow and oxygen uptake of the cerebral cortex. Br J Anaesth 39:186–196
22. McDowall DG, Harper AM (1964) Cerebral blood flow during trichlorethylene anaesthesia: a comparison with halothane. Br J Anaesth 36:11–18
23. McDowall DG, Harper AM (1965) Blood flow and oxygen uptake of the cerebral cortex of the dog during anaesthesia with different volatile agents. Acta Neurol Scand [Suppl] 14:146–151
24. McDowall DG, Harper AM (1969) Cerebral oxygen uptake and cerebral blood flow during the action of certain anaesthetic drugs. In: Betz E, Wüllenweber R (eds) Pharmakologie der lokalen Gehirndurchblutung. Ärztl Forsch Sonderbd, p 108–110
25. McDowall DG, Jacobsen J (1963) Cerebral blood flow during halothane anaesthesia. Br J Anaesth 35:394–402
26. Martin JT, Faulconer A, Bickford RG (1959) Electroencephalography in anesthesiology. Anaesthesiology 20:359–376
27. Neigh JL, Garman JK, Harp JR (1971) The electroencephalographic pattern during anesthesia with ethrane. Anaesthesiology 35, 5:482–487
28. Opitz A, Degen R, Oberwetter WD, Degen HE (1980) Inhalationsanästhesie bei Patienten mit zerebralen Krampfanfällen. In: Opitz A, Degen R (eds) Anästhesie bei zerebralen Krampfanfällen und Intensivtherapie des Status epilepticus. Perimed, Erlangen, p 65–75
29. Pearcy WC, Knott JR, Bjurstrom RO (1957) Studies on nitrous oxide, meperidine and levallorphan with unipolar electroencephalography. Anesthesiology 18:310–315
30. Pearcy WG, Knott JR, Pittinger CB, Keasling HH (1957) Electroencephalographic and circulatory effects of chloroform-anesthesia in dog. Anesthesiology 18:88–96
31. Pichlmayr I (1969) Über den Einfluß verschiedener Narkosearten auf Durchblutung und Funktion der Leber sowie Durchblutung der Hirnrinde. Habilitationsschrift, München 1967, Summarized in Fortschr Med 87:47–48
32. Pichlmayr I (1969) Das Verhalten der Hirndurchblutung bei Hunden unter verschiedenen Narkosearten. Z Kreislaufforsch 6:662–676
33. Pichlmayr I (1971) Die Bedeutung zerebraler Kreislaufveränderungen in Narkose. Fortschr Med 89:1087–1089
34. Pichlmayr I, Eichenlaub D, Keil-Kuri E, Klemm J (1970) Veränderungen der Hirndurchblutung unter Thiopental, Halothan und Fentanyl-Droperidol. Anaesthesist 19:202–204
35. Pichlmayr I, Lips U (1980) Halothane-Effekte im Elektroenzephalogramm. Anaesthesist 29:530–538
36. Possati S, Faulconer A, Bickford RG, Hunter RC (1953) Electroencephalographic patterns during anesthesia with cyclopropan. Correlation with cyclopropan in arterial blood. Anesth Analg 32:130–135
37. Schmahl FW (1965) Effects of anaesthetics on regional content of some metabolites of the brain cortex of the cat. Acta Neurol Scand [Suppl] 14:156–159
38. Stockard J, Bickford R (1974) The neurophysiology of anaesthesia. In: Gordon E (ed) A basis and practice of neuroanesthesia. Excerpta Medica, Amsterdam
39. Uyüc A (1954) Die Kontrolle der Narkosetiefe mit dem Elektroenzephalogramm. Anaesthesist 3:69–73
40. Yurewich WM (1974) A clinical and encephalographic comparison in the design of equipment for indicating degree of anaesthesia with methoxyflurane. Biomed Eng 7:145–147

IV. Intravenous Anesthetics

CONTENTS

The intravenous induction of anesthesia has the great advantage of a rapid onset without any significant psychic distress to the patient. In 1847, the infusion of ether was used for the first time of intravenous anesthesia by Pirogoff. Worldwide acceptance of intravenous anesthesia was brought about by Weese and Scharpf in 1932, who developed the water-soluble barbituric acid derivatives [64]. A large number of substances have since been developed and introduced for intravenous use. They differ according to the degree of distress they induce in parenchymous organs and their ability to be regulated; they can be used selectively for anesthesia. A classification of the various intravenous drugs assists in providing a summary of the choices available and in indicating drug application. The chemical substances for i.v. use can be divided into two major groups according to their historical development – the barbituric acid derivatives (barbiturates) and the non barbituric acid anesthetics. The latter are a mixed group as they include some drugs that are chemically unrelated. KUGLER (1981) suggested a classification of the intravenous hypnotics and anesthetics into five groups on the basis of their central nervous site of action. They are listed below.

1. Holencephalic Anesthetics. This group encompasses substances that paralyze the functions of extensive parts of the telencephalon, diencephalon, and mesencephalon after a brief period of activation – thus inducing a loss of consciousness (neocortical sleep). By decreased activity in the thalamus and limbic system, affects and pain perception are diminished. Overdose results in a blockage of the vegetative brain stem nuclei, endangering the patient through respiratory arrest and circulatory dysregulation. Barbiturates are considered typical representatives of this group.

2. Telendiencephalic Anesthetics. These are substances that depress only the cortical functions and pain processing structures in the thalamus, resulting in the so-called dissociative anesthesia of lost consciousness and analgesia. The upper brain stem, the limbic system, and other thalamic areas remain intact, or they are disinhibited or even increase in activity. Nightmares or psychomotor activities may occur as side effects. Ketamine is typical of this pharmacological group.

3. Telencephalic Hypnotics. Drugs of this group mainly limit cortical activity. Deeper parts of the central nervous system remain primarily unaffected, but they may develop increased activity due to the loss of cortical inhibitive mechanisms. Clinically, this is manifest as motor reactions (myoclonias). The drugs, of which etomidate is a characteristic example, are pure hypnotics without any analgetic component.

4. Dienmesencephalic Neuroleptics. The main representative of this group is droperidol. By suppression of dien- and mesencephalic structures and of the limbic system, pain perception, emotions, and affectivity are diminished in a dose-dependent fashion.

5. Peripheral and Dienmesencephalic Analgetics. The common analgetics influence both the peripheral pain conduction and the thalamic structures. Apart from the clinically desired pain relief, they result in a general subduing of consciousness, due to additional effects on diencephalic areas, the limbic system, and the cortex. Representative of this group is fentanyl.

Although the cerebral site of action allows an easy classification of newly developed anesthetics, hypnotics, and psychotropic drugs, it has not been accepted in anesthesiology, where the classification according to chemical group is still used. Thus, for practical purposes in anesthesiology, we are going to follow the distinction according to pharmacological criteria.

Monoanesthetics

Barbiturates

Pharmacological Mode of Action and Clinical Application. Barbiturates are used worldwide mainly for the intravenous induction of anesthesia. At present, of the many barbituric acid derivatives that have been developed since the first reports on the successful application of hexobarbital [64], the following are considered anesthesiological standard drugs: hexobarbital, introduced in 1932 [64]; thiopental, introduced in 1953 [43]; and methohexital, introduced in 1957 [62]. Most studies have been done on thiopental, which is used as a standard drug for comparison with newly developed anesthetics (Benzer et al. 1982). After intravenous injection, the pharmacokinetics of all barbituric acid derivatives is similar. The brief narcotic effect is due to a process of redistribution. Depending on central and peripheral perfusion, lipid solubility, plasma-protein-binding, and the degree of dissociation, the intravenously administered barbiturate reaches the brain (55%), heart, splanchnic area, and kidneys after approx. 1 min [54, 55]. With free passage through the blood-brain barrier, the brain tissue level and the plasma level of barbiturates rapidly decrease by further migration of the substance into muscle and other less-perfused tissues. The resulting anesthesia lasts about 7–10 min depending on the dose. The metabolic gradation processes are slow, however, as is shown by the biotransformation of 10–15%/h for thiopental [52] and 10–19%/h for methohexital [44]. Barbiturates in therapeutic doses diminish cardiac output; with overdose, there is a depression of the central respiratory and circulatory regulation. Despite being generally referred to as "ultrashort barbiturates," hangovers occur long after the actual end of anesthesia, preventing the patient from driving [16]. Barbiturates cause a dose-related decrease in cerebral perfusion, cerebral metabolism, and CSF pressure (Benzer et al. 1982; Nemes et al. 1979).

Hexobarbital (Evipan)

Pharmacological Mode of Action and Clinical Application. The N-methylated barbiturate hexobarbital is used as a long-acting barbituric acid derivative in anesthesiology. The unaltered substance is excreted renally after 8 h after having been metabolized in the liver. The metabolic products, which are 80% hypnotically ineffective, are also excreted renally within 25 h. Hexobarbital is at present only used in special indications, mainly in intensive care units, due to difficulties in its regulation.

EEG Findings

Our Findings (32 patients, 16 patients under 50 years of age, 16 patients over 70 years; dose 7 or 14 mg/kg Evipan i.v. to each of 16 patients for in-

Table 1. Changes in the EEG background activity after intravenous application of hexobarbital ($n = 32$)

Frequency range	EEG power change after medication	Incidence (%)		Change in voltage	Dose of hexobarbital
Alpha	↑ Dominance 5–10 min	69	Occurrence of an irregular EEG 38%	20– 75 µV	7 mg/kg
Beta	↓ Dominance 5–10 min				
Delta	↑ Dominance 1– 5 min	62		100–200 µV	
Theta	↑ Dominance 1– 5 min				
Burst-suppression phases[a]	0	0		0	
Alpha	↑ Dominance 5–10 min	69	Occurrence of an irregular EEG 31%	20– 75 µV	14 mg/kg
Beta	↑ Dominance 5–10 min				
Delta	↑ Dominance 1– 5 min	56		100–200 µV	
Theta	↑ Dominance 1– 5 min				
Burst-suppression phases	+	63		0	

[a] 0, absence; +, presence

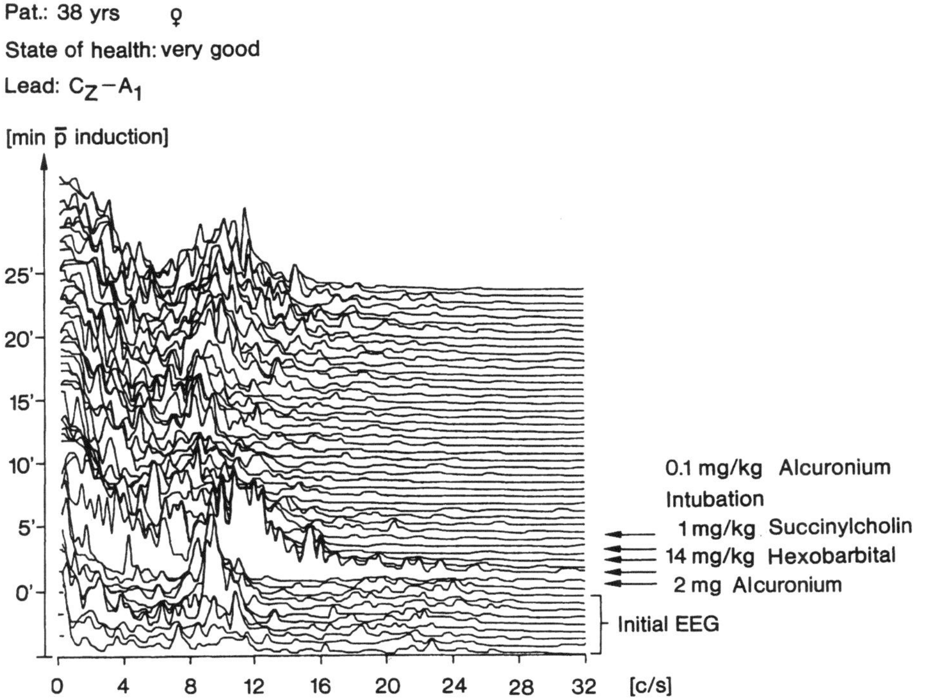

Fig. 1. After i. v. hexobarbital administration, following a desynchronization period lasting approximately 1 min, a pure delta EEG is formed, which gradually accelerates; toward the end of the observation period it shows a typical irregular EEG of 0.5–16 Hz with a dominant frequency in the theta and alpha ranges. This EEG course, which also corresponds with the clinical stages of anesthesia, is typical for all barbiturates.
Recording conditions: TC, 0.3 s; filter, 70 Hz; calibration, 50 µV = 7 mm; FFT, 30-s periods

duction of anesthesia) (Table 1; Fig. 1). The induction of anesthesia with hexobarbital, injected within 1 min, leads, independent of the initial EEG pattern, to irregular high-voltage EEG activity between 1.5 and about 25 Hz. Dominant activities of approximately 100 and 200 μV initially (for 3–5 min) lie in the delta/theta range; between the 5th and 10th min they move toward the alpha and low beta ranges. Whereas the first phase of the EEG changes proceeds in a relatively uniform manner, the second phase shows large interindividual differences in the frequency distribution of the alpha/beta band and also when returning to the dominant initial frequency. With the high dose of 14 mg/kg, the phase of synchronization persists somewhat longer. Fast frequencies are occasionally absent. A special development, labeled "overdose phenomenon," is seen especially in EEG recordings of geriatric patients: after a short period of synchronization, there is a depression of the electric activity, characterized by isoelectric phases and interrupted by single disordered discharge groups (bursts).

Clinical Evaluation. The administration of 7 mg/kg of hexobarbital in all patients was insufficient for induction of anesthesia. Intubation carried out with the patient relaxed caused defensive movements, increase in blood pressure, and tachycardia. Of the 16 patients receiving 14 mg/kg hexobarbital, ten patients (62,5%) were sufficiently anesthetized and tolerated the intubation without motor or vegetative reactions.

Interpretation of EEG Changes and Clinical Findings. The depth of anesthesia with hexobarbital can be clearly determined by spectral analysis: The high-voltage delta/theta activity indicates whether the depth of anesthesia is sufficient. With insufficient anesthesia, there is activation in the beta band. From the transition of synchronization (decrease in frequency, increase in amplitude) to desynchronization (increase in frequency, decrease in amplitude), one can actually observe recovery from anesthesia prior to clinical signs of wakening. Overdose may be diagnosed from the EEG before clinical signs of vegetative depression occur if burst-suppression phases are seen.

Thiopental (Trapanal, Pentothal)

Pharmacological Mode of Action and Clinical Application. At present, thiopental is the standard drug among the intravenous anesthetics. Its narcotic potential is double that of hexobarbital. Thiobarbiturates chemically form a separate class of barbituric acid derivatives. Metabolism proceeds slowly, but is almost completely in the liver. The metabolized products are excreted renally over a period of up to 7 days. Anesthesia is deepest 20–40 s after intravenous injection and becomes progressively lighter up to awakening, after 10–30 min.

EEG Findings

Our Findings (50 patients, 20–80 years of age, dose 7 mg/kg Trapanal i.v. over 1 min) (Table 2; Fig. 2). Thiopental provokes encephalographic changes similar to those of hexobarbital independently of the initial EEG. Obvious differences in the degree of the EEG changes, and thus of the

Table 2. Changes in the EEG background activity after i.v. application of thiopental ($n=50$)

Frequency range	EEG power change after medication	Incidence	Change in voltage	Dose of thiopental
Alpha	a) / b) ↑ Dominant ↑ 5.46 ±0.50 min	All patients – a) 52%, b) 48%	Approximately 20–75 μV	7 mg/kg
Beta		Different patterns in patients over 50 years (n=26) a) 89% b) 12%	Approximately 50–200 μV	
Delta, Theta	↑ Dominant 7.44±0.32 min			
Burst-suppression phases[b]	+	30% in patients above 50 years – 46%	0	

After the initial changes mentioned here, similar patterns were seen in all patients – a gradual shifting of frequency into the beta range, with loss of delta/theta activity

[b] 0, absence; +, presence

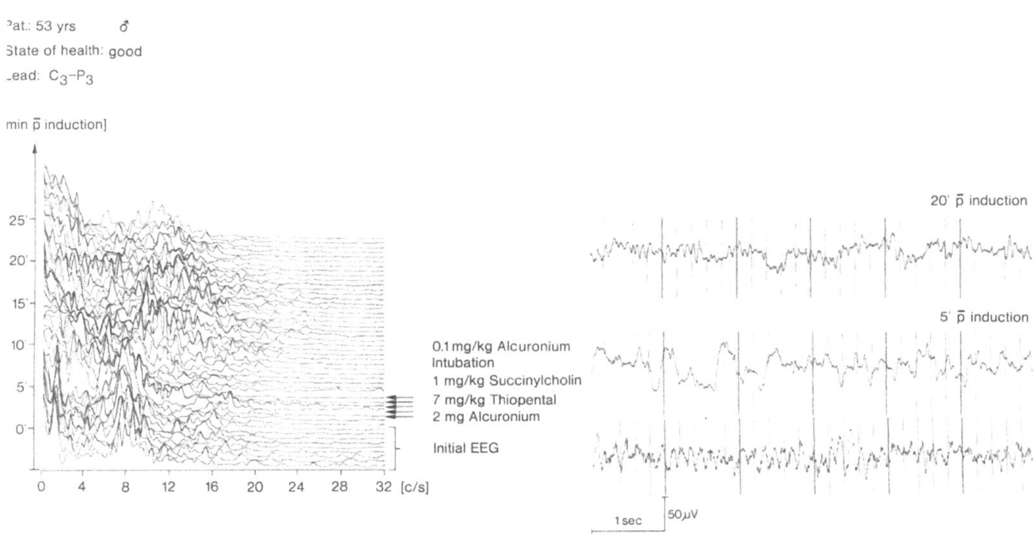

Fig. 2. Typical EEG changes after induction with thiopental. An initial synchronization period lasting 4–5 min is followed by a smooth transition toward desynchronization, indicating lighter anesthesia. The frequency pattern of the initial EEG during the observation period does not return.
Recording conditions: TC, 0.3 s; filter, 70 Hz; calibration, 50 μV=7 mm; paper speed, 30 mm/s; FFT, 30-s periods

depth of anesthesia under weight-related doses, are detected between patients under and over 50 years of age. In younger patients (n = 17 below 50 years), about 30 s after i.v. injection, the formation of a high-voltage irregular EEG can be seen with frequencies ranging along all four bands. During the first 5–6 min, the delta/theta range remains dominant. Then, as anesthesia becomes lighter, the dominant frequency range shifts into the alpha or beta band. This frequency pattern of light anesthesia is maintained with thiopental for 25 min. This pattern is seen in only 9.1% of patients over 50 years of age (n = 33). The remainder, immediately after i.v. thiopental administration, show a pure synchronization of the EEG with high-voltage 2–8 Hz waves without beta activity. This lasts for 7–8 min and then becomes non synchronous with high beta frequencies. In 15 patients, suppression phases were seen as a sign of excessive anesthesia (three patients under 50 years of age, 12 patients over 50 years).

Clinical Evaluation. With the thiopental dose mentioned above, a depth of anesthesia is obtained sufficient for intubation in 38 patients, but in 23 patients this soon woce off and prevented further manipulations. During the observation period of 25 min, 15 patients remained in deep anesthesia. In 12 patients the cerebral suppression was insufficient. Patients over 50 years "sleep" longer and deeper than younger ones.

Interpretation of EEG Changes and Clinical Findings. Delta/theta dominance in the EEG with thiopental indicates a sufficiently deep anesthesia, whereas alpha/beta dominance is a sign of light or excessively light anesthesia. Transient individual overdoses are detectable by burst-suppression phenomena in the EEG, even when clinical symptoms are absent.

Methohexital (Brevimytal, Brietal)

Pharmacological Mode of Action and Clinical Application. Methohexital is an N-methylated barbiturate chemically related to hexobarbital. The sulfur in the molecule is absent. It is a short-acting substance ("ultrashort barbiturate"); as with all barbiturates, this is due to redistributive processes. In addition, the rate of metabolism is slightly higher than that of comparable substances. The breaking down process takes place through hydrolization of the side chain; excretion of the metabolic products is renal. The narcotic potential is three-fold that of thiopental.

EEG Findings

Our Findings. Methohexital has only been used in our own patients in combined anesthesia. Since overlapping effects of the drug occur in the EEG, the results will be discussed below (p. 113).

Clinical Evaluation

With a weight-related dose of 1 mg/kg, an anesthetic depth sufficient for intubation was observed in 66.7% of the 39 cases. In the remainder, 33.3%, there were defensive movements with coughing or gagging and sometimes tachycardia. Unlike with thiopental, geriatric patients present the same distribution pattern as the younger patients.

Evaluation of Barbiturate-Induced EEG Changes (Table 3)

Electroencephalographic changes with barbiturates have been thoroughly studied and their course is well known [2, 4, 20, 45]. Barbituric acid derivatives generally lead to similar phenomena; dose, rapidity of entering the brain, and mode of excretion of the substance determine the pattern and degree of EEG stages of anesthesia. In combined anesthesia with nitrous oxide/oxygen and muscle-relaxants, the barbiturate effects determine the EEG pattern. With slow continuous barbiturate injection, the stages of anesthesia as described in Chap. BII, are passed through systematically. Initially, when somnolence is clinically evident, fast high waves occur, frequently with the formation of spindles. With the onset of sleep and loss of consciousness, high-voltage slow frequencies join the picture. The increased depth of anesthesia is indicated by this synchronization. The fraction of additional fast waves becomes progresssively smaller. Overdose with slow waves in the EEG, first become more level due to cerebral depression, then burst-suppression phases develop and finally there are long-lasting isoelectric periods. With rapid injection, very deep stages of anesthesia are immediately achieved, which are quickly lost when the substance is redistributed.

'able 3. EEG changes due to barbituric acid derivatives during the induction and subsequent stages of anesthesia ·eview of literature and our own findings)

:EG changes after barbiturates according to classification of anesthetic stages by Kugler 1981						
$_0$ A_1 A_2	B_0 B_1 B_2	C_0 C_1 C_2	D_0 D_1 D_2	E_0 E_1 E_2	F	Source
–13 Hz	13–25 Hz → Occasional spindles, high amplitude, approx. 100 µV	10–25 Hz → DF 14–15 Hz 75–100 µV	4–8 Hz Occasional fast waves, approx. 150 µV	1–2 Hz approx. 75 µV 1–2 Hz → 20 µV	BS	[2, 7, 16, 18, 30, 31, 61] Brechner and Walter 1962; Kugler 1981; Sadove et al. 1967
	← Wide fluctuations →					
	2–25 Hz DF 15 Hz → DF 8 Hz 100–200 µV		DF 5 Hz 150 µV	DF 2–4 Hz 100–40 µV	BS	Our results

\bbreviations: DF, dominant frequency; BS, phases of burst suppression

The rate of metabolism determines the disappearance of the barbiturate effects in the EEG. The EEG changes of anesthesia are then passed through in reverse order. After the barbiturate-induced EEG changes have subsided and there is a return to the dominant initial frequency, additional fast waves remain as a sign of barbiturate hangover, which exist postnarcotically for hours or days in the conscious patient.

Summary of EEG Findings with Barbiturates

With continuous EEG recording of barbiturate effects, findings characteristic for the stages of anesthesia quickly fluctuate between symptoms of light anesthesia and overdose. The EEG exactly indicates the actual depth of anesthesia; it is superior to clinical monitoring by the anesthesiologist. Because a reduced state of health, old age, or considerable concomittant ailments – even with a smaller weight-related dose – may provoke signs of overdose, transient stages of excessively deep cerebral depression are detectable only by EEG monitoring. For general adjustment and maintenance of a particular depth of anesthesia with barbiturate combination anesthesia, the continuous recording of the EEG parameters, especially in problematic anesthesia, is a considerable monitoring aid.

Nonbarbiturate Anesthetics

Etomidate (Hypnomidate)

Pharmacological Mode of Action and Clinical Application. Etomidate is a pure hypnotic without any analgetic properties. Clinically introduced in 1972 [15], it is suitable for inducing and maintaining anesthesia in relatively pain-free, short surgery. The selective effect on the cortex occurs without suppression of subcortical structures after 20 s. In addition to not providing analgesia, occasional clinical signs of increased activity from deeper structures of the brain can result with undesired motor signs, e.g., myoclonias [57]. The recommended dose for induction of anesthesia is 0.15–0.3 mg/kg. The hydrolytic metabolism of the substance occurs rapidly due to liver esterases; metabolites are excreted renally. Heart and circulation are hardly affected by etomidate; cerebral perfusion and intracranial pressure are reduced (Nemes et al. 1979).

EEG Findings

Our Findings (32 patients 20–80 years of age, dose 0.3 or 0.6 mg/kg to each of 16 patients) (Tables 4, 5; Fig. 3). Etomidate changes the EEG back-

Table 4. Changes in the EEG background activity after i.v. application of etomidate ($n = 16$); observation period 8 min

Frequency range	EEG power change after medication[a]	Incidence (%)	Change in voltage	Dose of etomidate
Alpha	↑ a) 1–4 min b) 1–8 min c) 4–8 min	a) 84 b) 9 c) 84	10– 50 μV	0.3 mg/kg
Beta	↑ 4–8 min	84	10– 30 μV	
Delta	↑ a) 1–4 min b) 1–8 min	a) 84 b) 9	50–100 μV	
Theta	↑ a) 1–4 min b) 1–8 min	a) 84 b) 9		
Burst-suppression phases	0	0	0	
Irregular = delta, theta/alpha/beta	Throughout the whole observation period	6.06	10–100 μV	

[a] ↑, increase; 0, absence

Table 5. Changes in the EEG background activity after i.v. application of etomidate ($n = 16$); observation period 8 min

Frequency range	EEG power change after medication[a]	Incidence (%)	Change in voltage	Dose of etomidate
Alpha	↑ a) 1–5 min b) –8 min	a) 81 b) 19	10– 50 μV	0.6 mg/kg
Beta	0	0	0	
Delta	↑ a) 1–5 min b) –8 min	a) 81 b) 19	50–100 μV	
Theta	↑ a) 1–5 min b) –8 min	a) 81 b) 19		
Burst-suppression phases	+	25 → 13 <50 yrs 25 → 37 >70 yrs		
Irregular = Delta, theta/alpha/beta	↑ 4–5 min	100	10–100 μV	

[a] ↑, increase; 0, absence

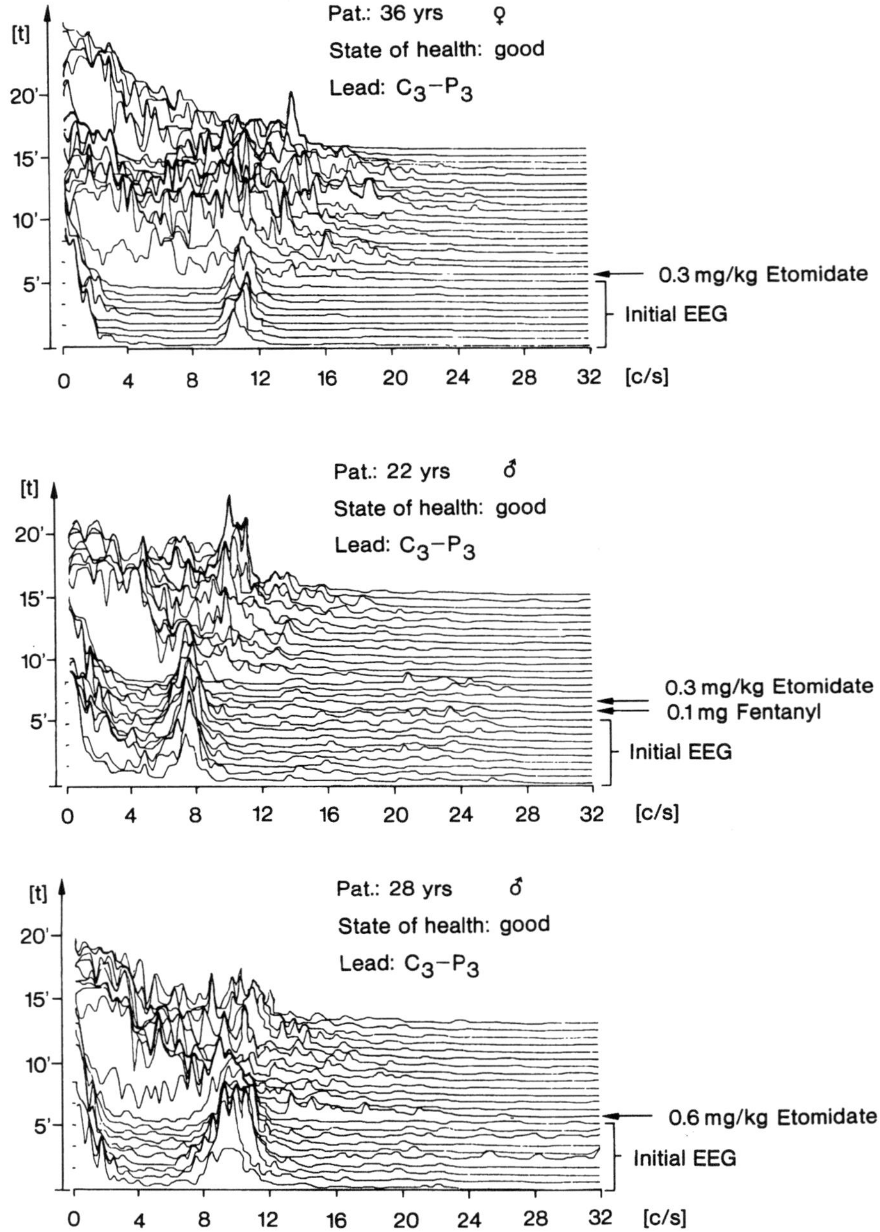

Fig. 3. Comparison of the EEG changes after various intravenous etomidate doses (patients under 50 years of age). All three examples demonstrate EEG changes similar to those seen with barbiturates as a result of the etomidate injection: Replacement of the initial activity by high-voltage delta/theta waves participation of the beta band at 0.3 mg/kg dose. The addition of fentanyl reveals less beta activity. After 10 min, a normalizing trend with reactivation of the initial situation is visible.
Recording conditions: TC, 0.3 s; filter, 70 Hz; calibration, 50 μV = 7 mm; paper speed, 30 mm/s; FFT, 30-s periods

ground activity toward an activation of all frequency bands. The delta/theta range is dominant during the first 4 min after injection; the initial alpha dominance returns between the 4th and 8th min with a still irregular frequency pattern. When doubling the dose, EEG signs (delta/theta dominance), corresponding with deeper suppression of consciousness, are maintained longer and are more intense. Beta waves are not seen. Burst-suppression phases indicative of overdose only occur after doses of 0.6 mg/kg bodyweight in 25% of cases mainly in geriatric patients.

Review of Other Findings. The EEG changes after i.v. administration of etomidate at 0.3 mg/kg are described as barbituratelike [41]. As in the induction phase with barbiturates diffuse irregular fast waves occur – as a sign of transient excitation of neuronal cortical activity – initially changing after some seconds into mixed activity of 3–5 Hz and 13–32 Hz – equivalent to light anesthesia. Moderate stages of anesthesia are characterized by higher slow waves with decreased fast activity. Initially, under deep anesthesia, high slow frequencies dominate. They gradually level off and in the case of overdose change into burst-suppression phases. During recovery from anesthesia, as usual the various stages are passed through in reverse order, with underlying slow waves being detectable over the posterior fossa, which is a particular feature of the recovery stage [14]. The postanesthetic phase is said to be characterized by wide changes in vigilance. Individual variance in the typical EEG changes is greater with lower doses.

Clinical Evaluation. The loss of consciousness caused by etomidate, usually lasts about 8 min. Manipulations on the patient, however, are only possible during the phase of delta/theta dominance. Myoclonias are more frequently seen with higher doses (25% in contrast to 10% with lower doses), younger patients, in particular, being more prone to this side effect (18.7% below 50 years of age, 6.3% above 70 years).

Interpretation of EEG Changes and Clinical Findings. From the reports in the literature and from our own findings it is obvious that etomidate may be included among those anesthetic substances whose narcotic effect and depth of anesthesia can be determined by EEG changes. As with barbiturates, the anesthetic stages 0–4 may occur according to dose. Because the clinical effect is brief, the transitions between the dominant frequencies are rapid.

Ketamine (Ketanest)

Pharmacological Mode of Action and Clinical Application. Ketamine, which is both intravenously and intramuscularly applicable, was discovered by Chen and Westen in 1959 [8], and was introduced for hospital use in

1964 by CORSSEN and DOMINO [9–11]. It has proven useful for all age-groups in short diagnostic or therapeutic surgery and in the induction of anesthesia, thanks to its anesthetic and analgetic properties. The therapeutic range for Ketanest is wide. Dose limits are set for intravenous application at 1.0–4.5 mg/kg, and for intramuscular application at 6.5–13.0 mg/kg. The effect occurs 30 s–7 min after injection. The hypnotic-analgetic duration of the telendiencephalically effective substance ("dissociative anesthesia") is about 10–15 min; the analgetic effect lasts 20–35 min longer. The duration of amnesia is 60 min. Nightmares have been reported as side effects – especially in younger adults [19, 31, 34, 35]. Quick distribution through the plasma to the tissues leads to peak plasma and tissue levels of the substance after 3–5 min; these levels are then progressively decreased by rapid metabolism. A second peak serum level 2 h after application indicates a possible return of the metabolites from the peripheral tissues. Serum half-life is about 15–20 min, the biological half-life approximately 4 h. Of the metabolized substance, 70% are excreted renally within 25 h, 83–96% within 5 days. The mode of action of ketamine differs from that of other anesthetics and involves the metabolism, the circulation, and cerebral changes. The stimulating effect on the general metabolism and circulatory functions, measured by an increase in oxygen input and oxygen consumption as well as by the blood pressure and pulse rate, are considered as central effects. On the one hand, they lead to stable or increased peripheral circulatory values, and on the other hand they result in contraindications in severe hypertension and coronary-risk patients. The increased metabolic rate can also cause a further increase in the oxygen debt under conditions of shock, and in combination with cardiac impairment it leads to a decreased ischemic tolerance in the organ affected. Ketamine, after a slight and brief decrease in the cerebral perfusion, leads to a considerable increase with raised intracerebral pressure. Cerebral oxygen consumption is also increased. EEG changes are evident for 120 min ([25] NEMES et al. 1979).

EEG Findings

Our Findings (89 patients 20–80 years of age, i.v. dose Ketanest: 1 mg/kg, n = 17; 2 mg/kg, n = 55; 3 mg/kg, n = 17) (Table 6; Fig. 4). After an intravenous ketamine dose of 1–3 mg/kg, typical changes in the EEG were evident. They were detected within 2 min and lasted for the observation period of 15 min. Characteristically, the disappearance of basic activity coincided with the occurrence of frequencies of about 4–7 Hz, with a potential difference of 40–90 μV. Additional activities that were already present were reduced. With some patients, especially the older ones, and with lower doses, the EEG changes were seen somewhat later (2 min postinjection). Frequently a more or less limited or persistent activity of high beta waves (28–32 Hz) was recorded at the same time. Occasionally, a premature return of the basic activity could be seen, in which case the dominant frequency was higher than it was originally (12–15 Hz).

Review of Other Findings. The EEG changes under intravenous ketamine administration observed by other study groups [12, 40, 56] also show the special properties of cerebral action compared with other kinds of anesthesia. Characteristically, in the stage of anesthesia, a regular theta activity of 4–6 Hz with amplitudes of 50 μV is seen bilaterally over the whole convexity, becoming lower and faster over a period of 20–40 min. It remains unaffected by psychosensory stimuli. In addition, occasional fast wave complexes and frequencies of 30–40 Hz occur at regular intervals. They are regarded by SCHWARTZ et al. [56] as a second characteristic of ketamine anesthesia. The complete pattern is replaced in the subsequent stage by long-lasting low basic activity with late alpha frequencies. Typically at this time, the clinically significant nightmares occur. Visible reactions to sensory stimuli remain minimal. The results are interpreted as follows: During the stage of surgical anesthesia, the cortical activity is considerably reduced (with little effect on rhinencephalic structures). The lack of cortical inhibition increases the activity of subcortical areas. A blockage of peripheral afferences then occurs at the site of origin of telencephalic projection pathways or in the periphery. In postnarcotic stage, the cortical output is still reduced. The activity of rhinencephalic areas is affected to a lesser degree, which is responsible for the spontaneous motor signs and for the intrapsychic events [40].

Clinical Evaluation. According to the depth of the loss of consciousness under ketamine, the percentage of very good effects increases from 23.5% (1 mg/kg) to 70.6% (3 mg/kg) with increasing doses. Light but sufficient anesthesia is, however, achieved just as well with low doses. Insufficient anesthesia occurs with 1 or 2 mg/kg in 29.4–34% of cases; with 3 mg/kg this occurs in 11.8% of cases. Within the test period of 15 min after ketamine ad-

Table 6. Changes in the EEG background activity after i.v. application of various doses of ketamines in patients with an initial alpha EEG

EEG	Dose			
	1 mg/kg ketamine (n = 17)	2 mg/kg ketamine (n = 55)	3 mg/kg ketamine (n = 17)	Independent of ketamine dose (n = 89)
Alpha	↓ 87%	↓ 78%	↓ 77%	↓ 77%
Beta	↑ 77%	↑ 60%	↑ 65%	↑ 64%
Delta	↑ 12%	↑ 16%	↑ 29%	↑ 18%
Theta	↑ 100%	↑ 96%	↑ 88%	↑ 96%

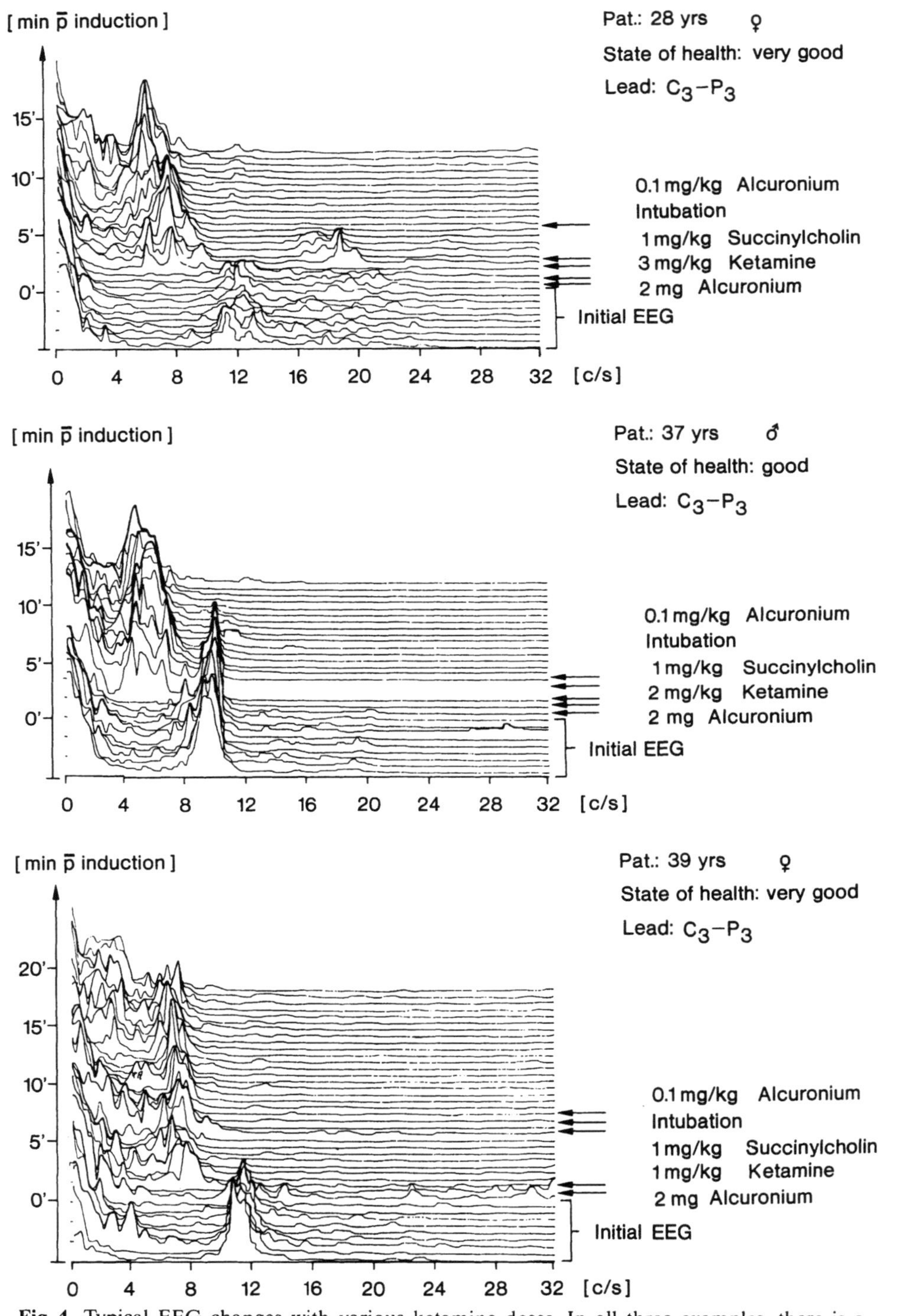

Fig. 4. Typical EEG changes with various ketamine doses. In all three examples, there is a transition of an initial alpha EEG toward the typical ketamine constant of 6 Hz (= theta EEG). Recording conditions: TC, 0.3 s; filter, 70 Hz; calibration, 50 μV = 7 mm; paper speed, 30 mm/s; FFT, 30-s periods

ministration, a change in the anesthetic depth sometimes took place, appearing more rapidly with lower doses. Correspondingly, the percentage of patients receiving 2 mg/kg ketamine with insufficient anesthesia was higher at the end of the test – 41.2%, compared with 11.8% receiving 3 mg/kg. Good and satisfactory depth of anesthesia was achieved at the end of the period of measurement with 1 mg/kg ketamine in 58.8% of cases, with 2 mg/kg in 61%, and with 3 mg/kg in 88.4%. Ketamine at any dose led to an increase in blood pressure and pulse rate over the whole period of observation (15 min), with peak values of 67% (blood pressure) and 45% (pulse rate) above the initial readings.

Interpretation of EEG Changes and Clinical Findings. Ketamine-induced EEG changes were first observed 1969 [12, 38]. Under optimal anesthesiological conditions, the basic activity is seen to disappear as are any additional frequencies and a characteristic power pattern is evident in the theta range of 4–6 Hz. The occasional occurrence of frequencies in the fast beta range (20–32 Hz), also observed in 1969, was seen in some of our own patients. Here, however, this may be attributed to lighter stages of anesthesia. From the pattern and course of the EEG changes, one can thus determine the degree of loss of consciousness due to ketamine.

Propanidid (Epontol)

Pharmacological Mode of Action and Clinical Application. Propanidid is an analgetically potent and short-term anesthetic. With a dose of 5–7 mg/kg, it takes effect – concomitant with deep hyperventilation – within 30 s and disappears quickly, having a half-life of 7 min. The substance is split hydrolytically into hypnotically ineffective metabolites by plasma cholinesterases and is eliminated from the body within 2 h. Propanidid is minimally toxic and would be ideal as a short-term anesthetic for outpatient surgery if it were not for its considerable circulatory effects and the drug-induced liberation of histamine, which has had fatal consequences and led to its being largely abolished. Propanidid leads to a decrease in cerebral perfusion and CSF pressure, lasting for almost 10 min ([53] Nemes et al. 1979).

EEG Findings. Owing to the potential hazards of propanidid we did not carry out our own studies.

Review of Findings in the Literature. With slow injection of the substance, the alpha rhythm of the initial EEG is at first disorganized; fast activities and sleep spindles appear, then the fast frequencies disappear in favor of slow delta/theta waves [17, 58]. The dominant slow waves parallel the decrease in cerebral oxygen consumption and cerebral perfusion, which is lowered by approximately 20–25% [63]. The EEG changes return to normal within 20 min (NEMES et al. 1979)

Clinical Evaluation. Propanidid is an exceptionally effective drug for short surgery. At present, severe potential complications rule out its clinical application. Upon recovery, a phase of slight euphoria is followed by full alertness within 2 h.

Interpretation of EEG Changes and Clinical Findings. With propanidid anesthesia, the EEG changes are similar to those with barbiturates. Postanesthetic effects of the substance do not appear to occur, in contrast to barbiturates, which show an increased incidence of beta frequencies for hours or days.

Althesine (Alfatesin, Althesin)

Pharmacological Mode of Action and Clinical Application. Althesine is a pure hypnotic with a wide therapeutic range. The recommended i. v. dose is 0.05 mg/kg. Loss of consciousness for 7–11 min occurs within 30 s, simultaneously with apnea. The substance is metabolized by hydrolytic splitting in the liver. Excretion of the metabolites mainly occurs with the fecies and urine. No metabolic products can usually be detected after 3 h. Moderate circulatory effects and the possibility of anaphylactic reactions must be taken into account in the application of the drug. With a transient cerebral decrease in metabolism of about 40%, the cerebral perfusion and intracranial pressure are correspondingly lowered. EEG changes vanish after 20 min ([13] NEMES et al. 1979).

EEG Findings. We did not carry out our own studies due to the possibility of anaphylactic reactions.

Review of Findings in the Literature. With a continuous increase in the depth of anesthesia by an increase in the dose, beta frequencies of between 10 and 24 Hz with a potential difference of 20–60 μV are initially activated. This beta rhythm then shifts into an alpha rhythm (8–12 Hz of 60–100 μV), with some residual fast and low beta activity. A stage of transient 4–5 Hz activation follows, changing into burst-suppression phases with further increase in the depth of anesthesia. Even under overdose, pain stimuli still provoke muscular defensive movements [3]. With a single weight-related application of the substance, theta waves of high amplitude occur for 15–20 min in addition to the remaining alpha activity; afterward a clear pattern of alertness is once more established [59].

Clinical Evaluation. Althesine is used to a small extent for the induction of anesthesia in surgery of long duration. The lack of any analgetic effects and disturbing muscle movements it produces – even with deep sedation – rule out its use for short-term surgery.

Interpretation of EEG Changes and Clinical Findings. Althesine, like the steroid hydroxydion, which is no longer used, produces EEG changes similar to those with barbiturates, depending on the actual hypnotic efficiency [37, 60]. Thus, with these drugs the stages of anesthesia can be determined from the EEG.

Summary of EEG Findings with Nonbarbiturate Substances

With the exception of ketamine, nonbarbiturate substances, during the induction of anesthesia and during the hypnotic phase, cause EEG changes similar to those produced by barbiturates. As the drug begins to take effect, beta waves are activated, which as anesthesia becomes deeper change into high slow waves, and with overdose result in burst-suppression phenomena. In lighter stages of anesthesia, some basic alpha activity frequently remains. The postnarcotic EEG changes, according to the mode and rate of metabolism, appear to be considerably shorter than with barbiturates. During the active phase, ketamine produces regular theta activity – with loss of frequency in the alpha band and occasionally with additional very fast waves; both patterns are regarded as typical.

Combined Anesthesia

Neuroleptanalgesia (Fentanyl and Droperidol)

Pharmacological Mode of Action and Clinical Application. With the clinical introduction of neuroleptanalgesia (NLA) by DE CASTRO and MUNDELEER in 1959 [6, 7], a method of anesthesia was introduced that was different from previous forms of intravenous anesthesia. This method is based on the selective suppression of surgical pain; it is thus more effective for the patient in that it combines a well-controllable, highly effective analgetic with a central nervous vegetatively suppressive substance, rather than reducing consciousness and depressing all central nervous functions as in general anesthesia. Suitable substances for NLA were developed by JANSSEN [28] and have been improved ever since. Today, the analgetic component is the pethidine derivative fentanyl, the analgetic effect of which is approximately 100-fold higher than morphine. The butyrophenon derivative dehydrobenzperidol (DHB) is applied as the neuroleptic. With the classic application one obtains – with the patient remaining responsive – a state of psychic indifference ("mineralization") plus analgesia.

Variations of NLA are today considered standard anesthesiological methods ([24, 27, 42, 50] HENSCHEL 1966, 1967, 1970). Pure NLA, based only on the above-mentioned substances and using muscle relaxants and artificial respiration, has widely been replaced by combined methods of barbiturate induction and the use of nitrous oxide or other volatile anesthetics as aids in reducing consciousness and supplementing or enhancing the neuroleptic analgetic effect. Fentanyl, like all opiates a highly potential analgetic, causes respiratory depression. With a rapid onset of action, its peak effect is reached after 20–30 min and its half-life is 1–2 h. With repeated injections – as used in anesthesia – the substance accumulates. Fentanyl is largely metabolized in the liver; excretion is up to 67% renal and occurs within 4 days. Undesired hangovers can be treated with opiate antagonists (e.g., Naloxone [51]). In addition to neurolepsy, droperidol causes a blockage of alpha receptors, with vasodilatation and possibly a cataleptic state. Metabolism is hepatic, excretion renal. With a half-life of 2.2 h, psychomotor aftereffects are observed for about 10 h even with low doses (5 mg) ([32] DELAY 1959). As anesthesiological medication, NLA lowers the cerebral metabolism and perfusion and intracranial pressure [1, 26, 33]. The advantages of NLA are seen in low toxicity of the anesthetics, a relatively stable circulation – after volume replacement subsequent to the vasodilative effect of DHB, and a relatively pain-free postoperative recovery.

EEG Findings

Our Findings (45 patients; 20–80 years of age; dose – 0.25 mg/kg DHB, 0.01 mg/kg fentanyl, 1 mg/kg succinylcholine for intubation, 3: 1 $N_20/0_2$; fluids – 250 ml electrolyte solution, 500 ml colloid solution) (Table 7; Fig. 5). Over an observation period of 25 min during induction of NLA, the following changes are usually seen: Alpha activity is first suppressed, returning more markedly after 5–10 min in 60% of patients. Beta frequencies, caused by premedication and present in more than half of the patients, are also suppressed. High delta/theta waves persist in 80% of the patients during the entire induction phase and determine the EEG picture.

Review of Other Findings. The EEG changes under NLA, unlike those of other kinds of anesthesia, were observed shortly after the clinical introduction of the method by de CASTRO and MUNDELEER [7] and have since been confirmed by other study groups [5, 47–49]. Extensive results were obtained by KUBICKI et al. [36, 38, 39]. In the induction and maintenance of general anesthesia by progressive cerebral suppression, a reversible detachment of the ascending reticular system (ARS) from the cerebral cortex takes place.

Table 7. EEG pattern during induction of NLA

Frequency range	EEG energy change after medication	Incidence (%)	Change in voltage	NLA dose
Alpha	↓ a) Temporary or permanent suppression	82	40–50 μV	0.25 mg/kg DHB
	↑ b) Return after 5–10 min	60		0.01 mg/kg fentanyl
	c) None	11		
Beta	Changes only when beta activity already present (in 51.1 % of cases)			1 mg/kg succinylcholine for intubation
	↓ a) Temporary or permanent suppression	65	20–30 μV	N_2O/O_2 3:1 $n=45$
	↑ b) Return or enhancement	70		
	c) None			
Delta	↑ a) Increase in activity	82	100–200 μV	
	(↑) b) Brief occurrence	7		
Theta	↑ a) Increase in activity	80		
	(↑) b) Brief occurrence	11		
Low-voltage EEG	+	4	10 μV	

Abbreviation: DHB, dehydrobenzperidol

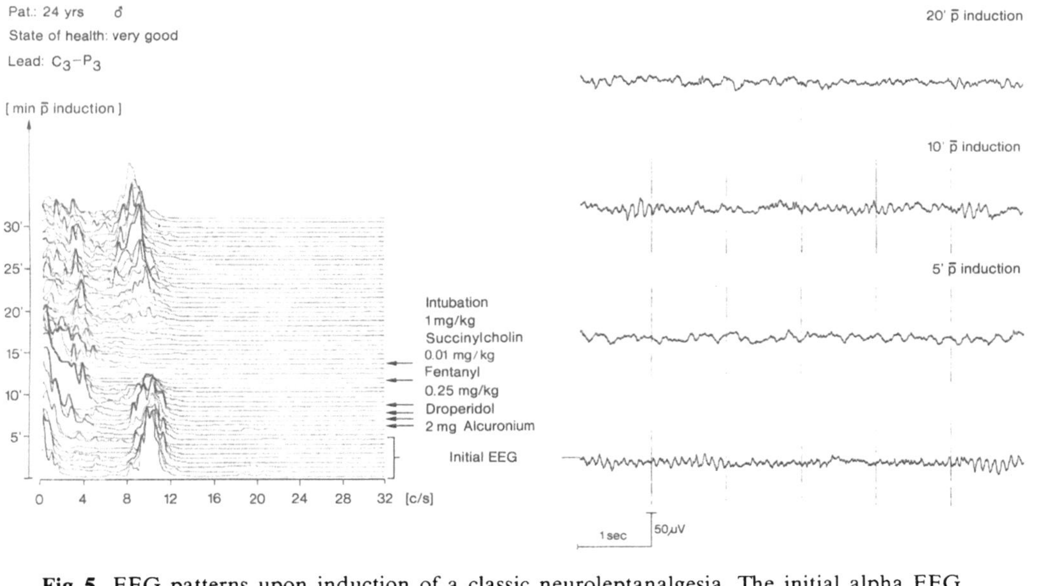

Fig. 5. EEG patterns upon induction of a classic neuroleptanalgesia. The initial alpha EEG disappears in favor of a high 0.5–4 Hz activity for approx. 8 min (= narcotic phase of the NLA). Apart from slow activity, high-voltage alpha waves return with a decelerated dominant frequency (= analgetic phase of the NLA).
Recording conditions: TC, 0.3 s; filter, 70 Hz; calibration, 50 μV = 7 mm; paper speed, 30 mm/s; FFT, 30-s periods

This is manifest in the encephalographically determinable stages of anesthesia and in the respective clinical signs of the depth of anesthesia. Under NLA, however, after a brief anesthesialike initial course, no detachment between ARS and cortex is thought to take place, but merely a shielding of higher cerebral areas by mesencephalic blockage of incoming impulses (Fig. 6). Accordingly, Kubicki classifies the NLA into a narcotic and an analgetic phase. With standard doses and application of the two neuroleptic analgetic components (droperidol and fentanyl), there is initially frequency deceleration down to the theta range, lasting approximately 15 min. This is said to be proof of the narcotically effective phase of the NLA. Although DHB, when applied alone, does not lead to a decrease in frequency, the cooperation of both substances and also the very rapid application of

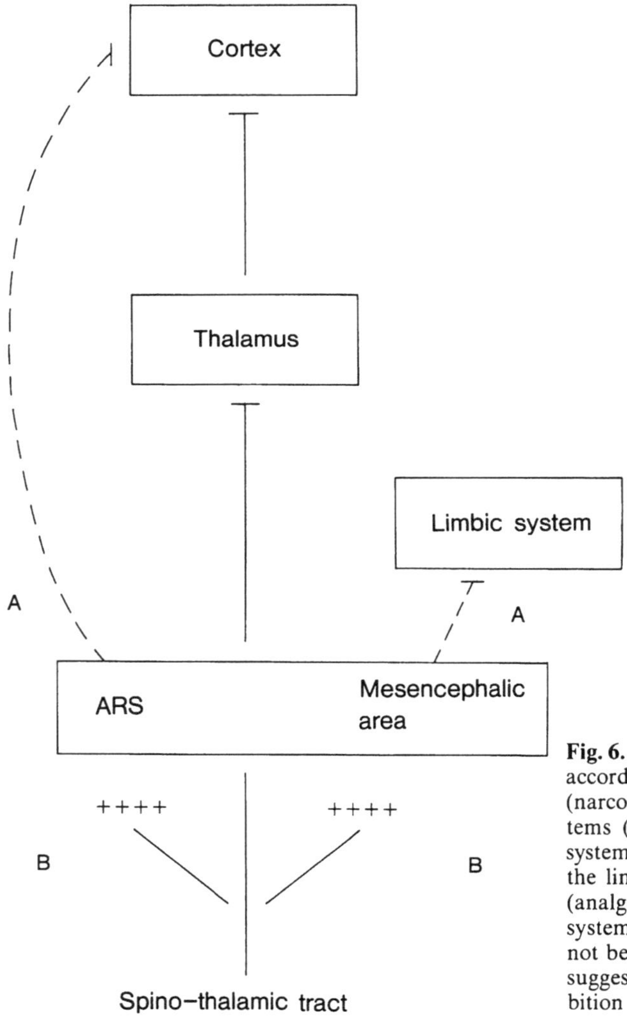

Fig. 6. Central mechanism of the NLA according to Kubicki et al. [38]. Phase *A* (narcotic phase), the marked control systems (- - -) from the ascending reticular system and from the mesencephalon to the limbic system are inhibited. Phase *B* (analgetic phase), the activity of these systems has returned to normal, but cannot be modified by peripheral influences, suggesting a sole synaptic connective inhibition (+ + + +)

fentanyl alone is responsible for the narcotic component of NLA. After 10–15 min, the alpha rhythm is resumed, characterizing the functioning ARS cortex. This alpha rhythm, in the sustained analgetic phase of the NLA, is extremely stable. It cannot be affected by peripheral stimuli. With repeated fentanyl doses for maintenance of the NLA, only slight alpha shifts into the upper theta range and no new "narcotic phases" are seen. After completing the NLA, following a discontinuation of surgical pain stimuli, a small decrease in frequency occurs once again, lasting up to 4 or even 6 h post operation. On the 1st postoperative day, patients after NLA show slower frequencies than after general anesthesia [21, 22].

Clinical Evaluation. With the dose mentioned above for NLA, short-term muscle relation, intubation, and subsequent respiration with nitrous oxide/oxygen, deep anesthesia is obtained in 33% of patients; 67% of patients show a defensive reaction during intubation or during other measures, such as applying a central venous pressure (CVP) line or central venous catheter.

Interpretation of EEG Changes and Clinical Findings. The clinical observations correlate well with the EEG findings during induction of NLA in our patients. A persistent alpha or beta EEG indicates light or excessively light anesthesia; a pure delta/theta EEG suggests sufficient anesthetic depth. The recurrence of alpha/beta activity coincides with clinical signs of recovery. Toward the end of the observation period, anesthesiological manipulations frequently initiate clinical arousal reactions with alpha/beta activation in the EEG. Thus, duration and depth of the controlled narcotic phase of NLA can be determined from the EEG findings.

Neuroleptanalgesia Induced by Other Intravenous Anesthetics

The relatively long period of induction in NLA and the awareness of intubation without amnesia are perceived by the patient as unpleasant, making the induction of anesthesia with other intravenous substances desirable, as it is widely practiced today. For this purpose, barbiturates are generally preferred.

Neuroleptanalgesia and Barbiturates

With barbiturate induction of NLA, barbituric acid derivatives are injected solely for bringing about the initial phase of sleep. Anesthesia is then continued as NLA with nitrous oxide/oxygen and muscle relaxants. In our own studies, the induction of NLA was tested with thiopental and methohexital, the EEG was recorded simultaneously.

Table 8. EEG pattern during NLA induced by a barbiturate (thiopental)

Frequence range	Type of EEG power change in barbiturate phase	NLA phase	Dose
Alpha	Typical barbiturate effects 100% [a]	↑ 73%	5 mg/kg thiopental 1 mg/kg succinylcholine 0.005 mg/kg fentanyl 0.15 mg/kg KG DHB Respiration with N_2O/O_2 3:1 $n=45$
Beta	4.7±0.7 min, continuous, then abrupt shift into the NLA phase in 93.3% of cases	↑ 27%	
Delta/theta		↑ 76% of which delta/theta without alpha = 2%	
Irregular = alpha/beta/delta/theta	After thiopental, irregular EEG occurs indistinguishable from the NLA phase, in 6.7% of cases	↑ 24%	

Abbreviation: DHB, dehydrobenzperidol
[a] See Table 2

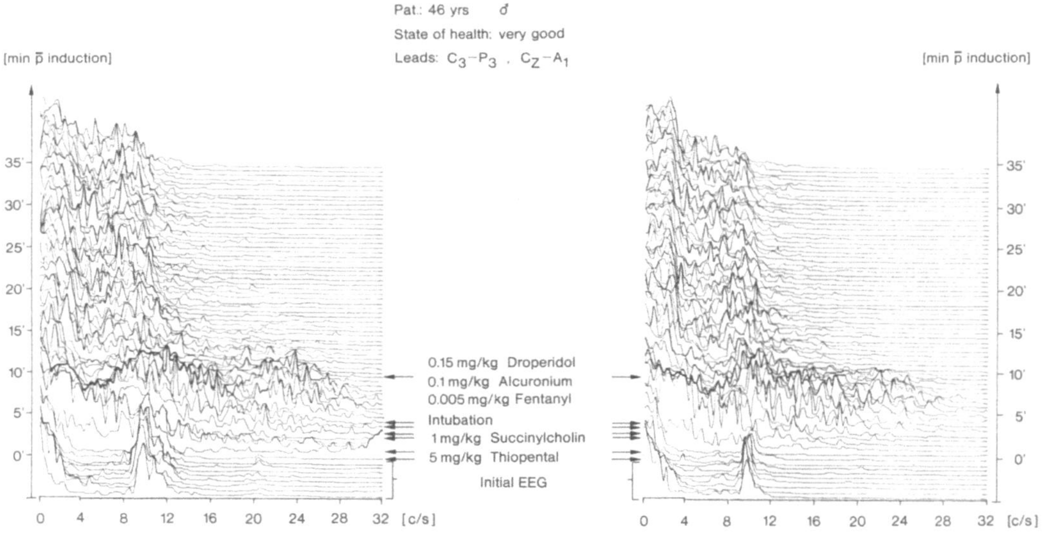

Fig. 7. EEG spectral analysis with thiopental-induced neuroleptanalgesia. The typical thiopental-induced EEG changes are abruptly converted, through the fentanyl injection, to the pattern of an analgetic phase of the NLA (delta/theta and alpha activities). The slowing of the dominant alpha frequency and the accentuation of the delta/theta fraction indicate a sufficient anesthetic depth.
Recording conditions: TC, 0.3 s; filter, 70 Hz; calibration, 50 μV = 7 mm; paper speed, 30 mm/s; FFT; 30-s periods

Table 9. EEG pattern during NLA induced by a barbiturate (methohexital) ($n=40$)

Frequency range	Kind of EEG power change in barbiturate phase	NLA phase	Drug dose
Alpha	Typical barbiturate effect in 100% of cases	↑ 70 %	1 mg/kg methohexital 1 mg/kg succinylcholine 0.005 mg/kg fentanyl 0.15 mg/kg DHB Respiration with N_2O/O_2, 3:1
Beta	Compare with table 2 (thiopental); 4.14 ± 1.1 min sustained, then	↑ 43%	
Delta/theta	abrupt change into NLA phase in 92.5% of cases	↑ 80%, of which only 10% delta/theta without alpha	
Irregular = alpha/beta/delta/theta	After methohexital, irregular EEG occurs indistinguishable from NLA phase in 7.5%	↑ 17%	

Abbreviation: DHB, dehydrobenzperidol

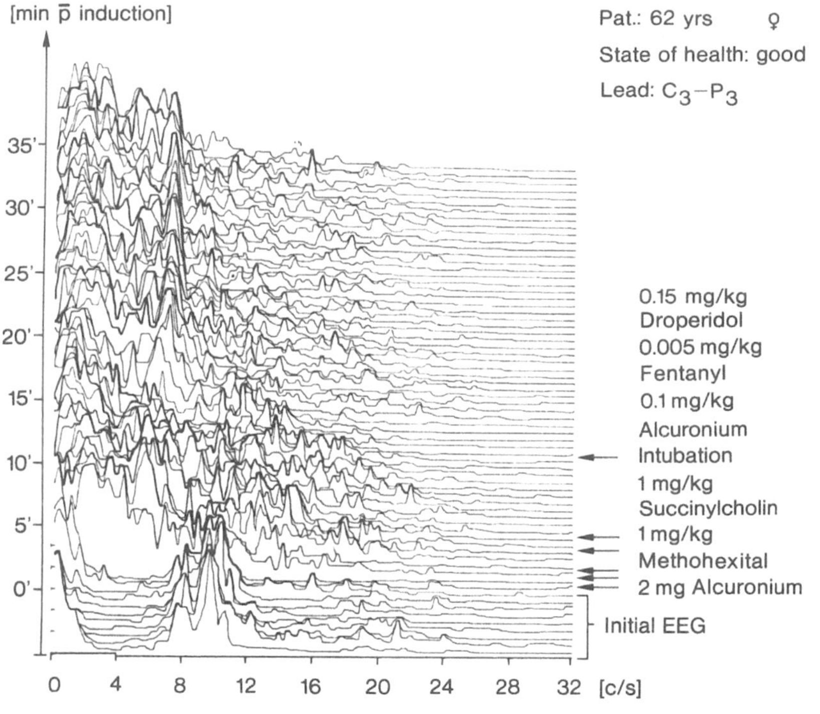

Fig. 8. EEG changes of a methohexital-induced neuroleptanalgesia. Methohexital initially leads to changes typical for the barbiturates. The fentanyl injection after intubation demonstrates the first "narcotic NLA phase," in contrast to Fig. 7. After approx. 7 min, the EEG pattern of the "analgetic phase" appears.
Recording conditions: TC, 0.3 s; filter, 70 Hz; calibration, 50 μV = 7 mm; paper speed, 30 mm/s; FFT, 30-s periods

EEG Findings

Our Findings (45 patients, 20–80 years of age, dose – 5 mg/kg thiopental, 0.15 mg/kg HDB, 0.005 mg/kg fentanyl; 40 patients, 20–80 years of age, dose 2 mg/kg methohexital, 0.15 mg/kg DHB, 0,005 mg/kg fentanyl) (Tables 8, 9; Figs. 7, 8). During barbiturate induction with *thiopental,* the typical EEG changes occur in an almost regular fashion as described above (p. 94). Irregular waves between 0.5 and 24 Hz with dominant frequencies in the delta/theta band characterize the barbiturate effects. Individual differences in the EEG changes are similar electrophysiologically and clinically to those seen with thiopental application alone. Shortly after the injection of fentanyl, abrupt changes in the EEG pattern are seen, with a frequent incidence of delta/theta activation and a simultaneous stable alpha rhythm. The alpha band sometimes shifts by about 1 Hz into the lower, occasionally into the higher, frequency range. Beta components may prevail, but indicate that anesthesia is too light. Subsequent addition of neuroleptics does not change the EEG pattern. The proportion of alpha activity is greater than with pure NLA. In geriatric patients with irregular initial EEGs, only the upper frequency limit is lowered by the substances tested. *Methohexital,* as drug for induction of NLA, leads to EEG changes similar to those described above for thiopental, both in type and extent. Despite the shorter action of the substance, the NLA phase begins when the barbiturate is taking effect and runs a parallel course.

Neuroleptanalgesia and Nonbarbiturate Substances

The hypnotic etomidate was used under EEG monitoring in our own patients both for induction of NLA and in combination with the analgetic fentanyl for pain reduction in surgery.

EEG Findings

Our Findings (33 patients, 20–80 years of age, dose – 0.3 mg/kg Hypnomidate, 0.005 mg/kg fentanyl, 0.15 mg/kg DHB; 20–80 years of age, 0.3 mg/kg Hypnomidate, 0.1 mg/kg fentanyl) (Tables 10, 11; Figs. 3, 9). During induction of NLA with the short-acting hypnotic *etomidate,* the EEG changes are initially determined by the hypnotic; after the injection of the analgetic/neuroleptic substances – administered shortly thereafter – these changes are again modified toward the EEG signs that are characteristic for NLA. The *combination* of the analgetic *fentanyl* with the hypnotic *etomidate* for short painful surgery shows different effects upon the cerebral function in geriatric and in younger patients. Basically, the EEG changes already described for both substances are found together. In younger patients, the course is characteristic and predictable to a limited extent. Initially, there is a 3-min activation of delta/theta frequencies, which remains constant over the tested period of 20 min in about 75% of cases except for the recurrence of alpha/beta activity with alpha dominance. In 25% of patients, the slow components disappear after the reestablishment of the alpha frequency with participation of the beta band. Only 12.5% of older patients react encephalographically in this way; in a further 12.5%, the EEG course is similar yet lacks the primary complete suppression of the alpha rhythm. Approximately 50% of patients over 70 years of age react with the formation of an irregular EEG, of which the upper frequency limit is only moderately indicative of the clinical anesthetic effect.

Table 10. EEG pattern during NLA induced by etomidate ($n=33$)

Frequency range	Kind of EEG power change in etomidate phase	NLA phase	Drug dose
Alpha	Typical etomidate effect[a]	↑ 76%	0.3 mg/kg etomidate
Beta	2.7±0.7 min sustained in 87.9% of cases	↑ 9%	1 mg/kg succinylcholine
Delta/theta		↑ 79%, of which only 3% delta/theta without alpha	0.15 mg/kg DHB 0.005 mg/kg fentanyl
Irregular = alpha/beta/ delta/theta	After etomidate, irregular EEG occurs indistinguishable from NLA phase in 12.1% of cases	↑ 21%	Respiration with N_2O/O_2, 3:1

[a] Tables 4, 5

Table 11. EEG pattern during intravenous application of fentanyl and 0.3 mg/kg etomidate ($n=16$)

Frequency range	EEG power change after medication	Incidence (%)	Change in voltage	Drug dose
Alpha	a) ↑[a] Dominant after 3 min	88	10– 50 μV	0.1 mg fentanyl 0.3 mg/kg etomidate
	b) ↑	12		
Beta	a) After 3 min	88	10– 30 μV	
	b)	12		
Delta/theta	a1) ↑ 1–3 min	25	50–100 μV	
	a2) ↑	75		
	b) ↑	25		
Irregular = alpha/beta/ delta/theta	a) ↑	12	10–100 μV	
	b) ↑	75		

[a] a., patients under 50 years of age; b., patients over 70 years

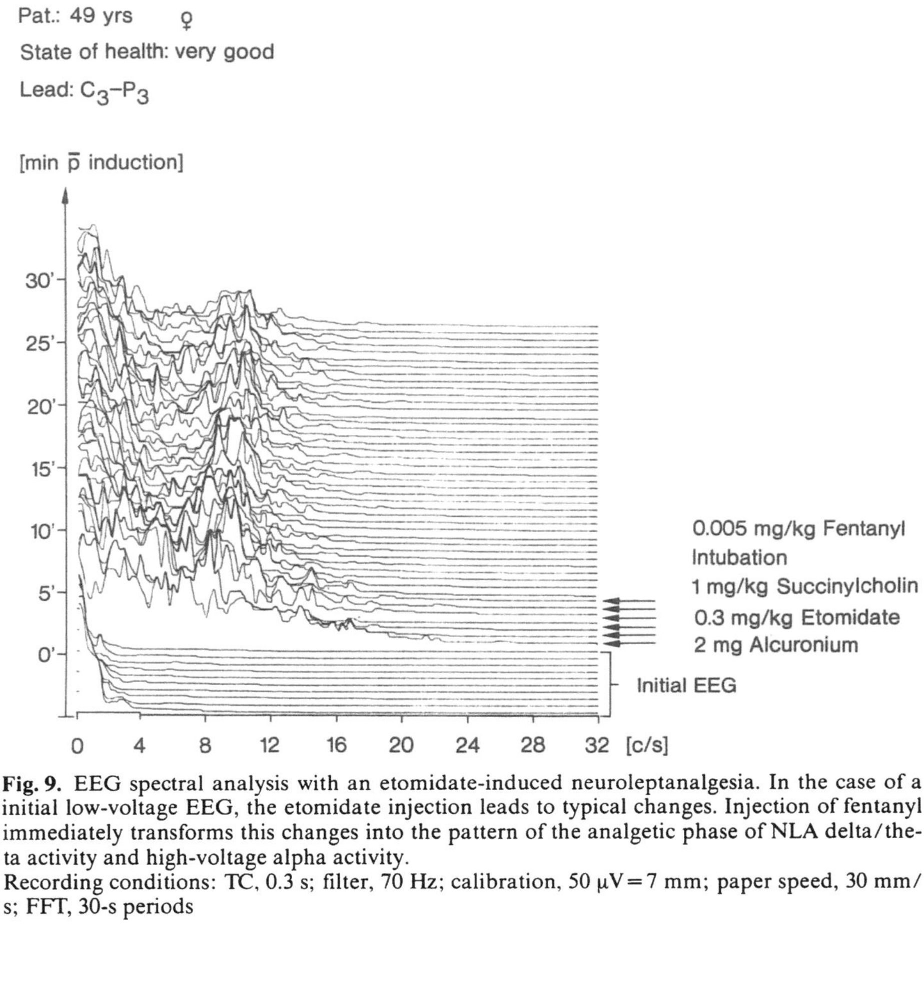

Fig. 9. EEG spectral analysis with an etomidate-induced neuroleptanalgesia. In the case of a initial low-voltage EEG, the etomidate injection leads to typical changes. Injection of fentanyl immediately transforms this changes into the pattern of the analgetic phase of NLA delta/theta activity and high-voltage alpha activity.
Recording conditions: TC, 0.3 s; filter, 70 Hz; calibration, 50 μV = 7 mm; paper speed, 30 mm/s; FFT, 30-s periods

Review of Other Findings. Modification of NLA with an additional narcotic for induction changes the EEG findings characteristic for NLA [40]. While the ARS function (cf. Fig. 6) is altered by conventional anesthesia and by the analgetic component of the NAL in different ways, these differences become less pronounced in all combination methods. The brain stem is generally inhibited both in the narcotic phase of the NLA and in barbiturate anesthesia. Striothalamically modulated sleep rhythms of the cortex with delta waves and beta spindles occur accordingly. This stage is more intense and lasts longer with barbiturate induction of NLA. The suppression of special brain stem functions acts by blockage of anatomic-functional structures and is responsible for the analgetic phase of the NLA, which causes a disconnection of afferent stimuli from the ARS and from the pain-processing centers in the limbic system. This is the analgetic phase of the NLA, it occurs later and is less defined. Correspondingly, a combination of barbiturates and

NLA provokes low frequencies in the case of a stable alpha rhythm, while nitrous oxide/oxygen in analgetic concentrations (3:1) does not alter the EEG characteristics of the NLA [27].

Clinical Evaluation. The induction phase of NLA is shortened by induction with intravenous narcotics, whereby over a relatively limited time the anesthesia remains at a deeper level than with NLA alone. In accordance with the different pharmacokinetics of the substances, etomidate acts more quickly than the barbiturates (2.75 ± 0,7 min as opposed to 4.4 ± 0.9 min, respectively). In clinical practice, great differences between young and geriatric patients are seen in the course of anesthesia with the combination of etomidate and fentanyl. Of the younger patients, 75% undergo sufficient anesthesia for short painful surgery – in accordance with the EEG findings; 25% are insufficiently anesthetized. Side effects, such as mild myoclonias, are rare (12.5%). In the geriatric patient group, where there is a generally lower narcotic potency of the combination (50% show good anesthetic effects), the frequency of side effects (75%) is markedly increased; 65.5% of these cases are due to brief respiratory arrest requiring assisted respiration.

Interpretation of EEG Findings with Induced Neuroleptanalgesia

With anesthesiological combination methods for induction of NLA, the EEG again clearly indicates the pattern of the anesthesia and its depth at any one time. The various cerebral sites of action can be determined by the individual phases of the EEG for each substance used for NLA induction. The EEG can be correlated with the clinical signs of anesthesia at all times. The drugs used for induction of NLA usually determine the EEG in the phase of falling asleep. EEG changes due to neuroleptanalgetic substances are seen after a few minutes and are maintained for a long time. A primary or drug-induced irregular EEG in geriatric patients rarely allows conclusions to be drawn regarding the prevailing anesthetic stage because anesthesia-induced changes (limitations of the upper frequency limit) only moderately correlate with the clinical effects.

Further Methods of Combined Anesthesia

Ketamine and Diazepam

Dissociated anesthesia induced by ketamine (see p. 101) may be maintained by repeated injections or continuous infusion of ketamine together with nitrous oxide/oxygen respiration and muscle relaxants during surgery of long duration. However, the incidence of vivid dreams and occasional psychic disturbances entail that this method has to be restricted to adults. A combination of ketamine with diazepam is, therefore, recommended [19, 29, 31, 34, 35, 46]. The sedative relaxing effect of the diazepam as well as its augmentation of muscle relaxation complements the hypnotic-analgetic ketamine effects during anesthesia; it also reduces dreaming and psychomimetic side effects postoperatively by long-term suppression of the limbic system and corpus amygdaloideum, with a consequent lowering of fear and aggression.

EEG Findings

Our Findings (45 patients, 20–80 years of age, dose – 250 mg ketamine/ 50 mg diazepam in 500 ml fructose) (Table 12; Fig. 10). The EEG under ketamine/diazepam – compared with the changes due to the single drug – with individual variability shows marked interference phenomena of both drugs. Encephalographicylly, the induction is characterized by the following features: The primary initial activities (mainly alpha frequencies) are suppressed with a latency of 4.1 ± 1.2 min after the beginning of infusion. Then, delta/theta activity is established as a manifestation of true anesthesia with the simultaneous appearance of high-voltage fast beta frequency for 5–10 min when it is clinically evident that the patient could be awoken, this is a sign of cerebral excitation, rather than a diazepam effect. Long-term EEG monitoring shows that with the infusion rate mentioned above, after 20–30 min, state of anesthesia is obtained that is sufficient for surgery; the delta/theta frequencies are the distinctive EEG features. Beta components are not seen at this time.

Table 12. EEG changes during i.v. ketamine/diazepam infusion with spontaneous respiration of room air ($n = 45$)

Frequency range	EEG power change after medication	Incidence	Degree of change	Drug dose
Alpha	↓	100	Changing voltage dependent on initial EEG over a wide range	250 mg ketamine and 50 mg diazepam in 500 ml electrolyte solution, 2 ml/kg initially; maximal drip speed until sleep, sleep maintained at 1 drop/kg min. Average total amount for 15-min anesthesia – 180.6 ml ± 36.0 ml = 90.3 ± 18 mg ketamine, 18.1 ± 3.6 mg diazepam
Beta	↑ 5–15 min	56		
Delta	↑ a) } ↑ c.	a) 33		
		b) 11		
Theta	↑ b)	c) 16		
Irregular = alpha/beta/delta/theta/	↑	11		
Low-voltage EEG		9	20 μV	

Latency of 4.1 ± 1.2 min until occurrence of first changes

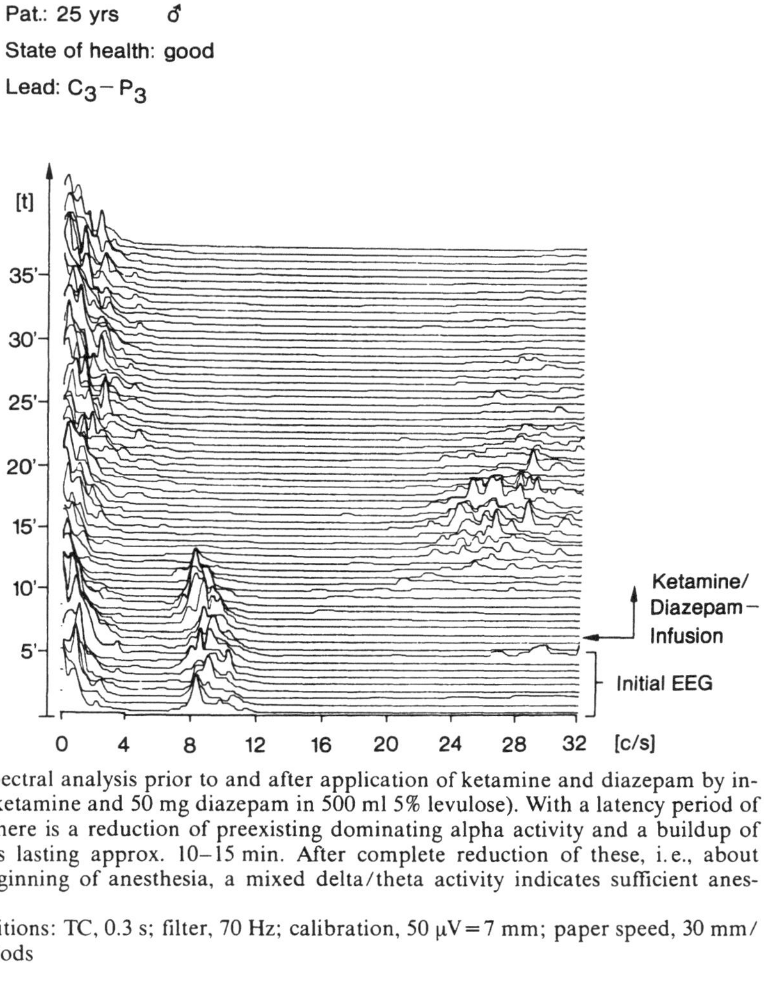

Fig. 10. EEG spectral analysis prior to and after application of ketamine and diazepam by infusion (250 mg ketamine and 50 mg diazepam in 500 ml 5% levulose). With a latency period of approx 7 min, there is a reduction of preexisting dominating alpha activity and a buildup of beta frequencies lasting approx. 10–15 min. After complete reduction of these, i.e., about 20 min after beginning of anesthesia, a mixed delta/theta activity indicates sufficient anesthetic depth.
Recording conditions: TC, 0.3 s; filter, 70 Hz; calibration, 50 μV = 7 mm; paper speed, 30 mm/s; FFT, 30-s periods

Review of Other Findings. HATANO et al. in their studies on more than 1000 patients undergoing ketamine/diazepam anesthesia by the Micro-Mini drip method carried out tests on selected patients (older patients or patients undergoing surgery with induced hypotension) [23]. In the discussion of their findings, they only emphasize the fast superimposed beta-spindle activity, which is characteristic for diazepam.

Clinical Evaluation. Ketamine-diazepam with no additional nitrous oxide or relaxant leads to an adequate state of anesthesia after 20–30 min. In some cases, however, this is not achieved. These drugs are, therefore, not suitable for short anesthesia. The weight-related dose is too rigid for carrying out anesthesia to meet individual requirements. To improve its clinical application, further substances are required, such as nitrous oxide and relaxants.

Interpretation of EEG Changes and Clinical Findings. The level of sedation and the degree of anesthesia at any one time under ketamine/diazepam can be derived from the EEG findings.

Summary of EEG Findings in Intravenous Anesthesia

The group of intravenous anesthetics comprises drugs that act cerebrally in different ways. As expected, they show some similarity during the induction phase, but also show characteristic differences in their action on the EEG pattern. Barbiturates and barbituratelike anesthetics or hypnotics give rise to the classic stages of anesthesia, as described in Chapter BII; nitrous oxide/oxygen or muscle relaxants do not influence the depth of anesthesia as documented by the EEG but lead to an improved evaluation by diminishing artefacts. NLA has its own EEG characteristics in the analgetic phase and features a stable unmodifiable alpha rhythm. The EEG changes in the first minutes of NLA ("narcotic phase") show typical delta/theta activities as an expression of the anesthesia and differ from the barbiturates by the absence of the initial excitatory activities (initial beta activation). It is notable that by the combined application of barbiturates in NLA, an injection of fentanyl abruptly suppresses any barbiturate effects and gives rise immediately to the pattern of the analgetic NLA phase. Of the nonbarbiturate substances, ketamine typically shows EEG features of high-voltage theta waves and, occasionally, activation of the upper beta range. The EEG monitoring of intravenous induction indicates the actual stage of anesthesia as well as the cerebral bioavailability of the substance in each case. Knowledge of the encephalographic characteristics of these substances is required for an evaluation of the EEG.

References

A. Textbooks and Summaries

Benzer H, Frey R, Hügin W, Mayrhofer O (1982) Lehrbuch der Anästhesiologie, Reanimation und Intensivtherapie, 5th edn. Springer, Berlin Heidelberg New York

Brechner W, Walter R (1962) Practical Electroencephalography for Anesthesiologists. Thomas, Springfield

Delay J (1959) Psychopharmacology Frontiers. Little/Brown, Boston

Henschel WF (1966) Die Neuroleptanalgesie. Springer, Berlin Heidelberg New York

Henschel WF (1967) Neuroleptanalgesie. Klinik und Fortschritte. Schattauer, Stuttgart

Henschel WF (1970) Neue klinische Aspekte der Neuroleptanalgesie unter besonderer Berücksichtigung methodischer Varianten. Schattauer, Stuttgart

Kugler J (1981) Elektroenzephalographie in Klinik und Praxis, 3rd edn. Thieme, Stuttgart
Nemes C, Niemer M, Noack G (1979) Datenbuch der Anästhesiologie. Fischer, Stuttgart
Pirogoff N (1847) Recherches practiques et physiologiques sur l'éthérisation. Imprimerie française, St Petersburg
Sadove MS, Becka D, Gibbs FA (1967) Electroencephalography for anesthesiologists and surgeons. Pitman, London

B. Papers

1. Baker J, Harper AM, McDowall DG, Fitsch F, Jennett WB (1968) Cerebral blood flow, cerebrospinal fluid pressure and EEG-activitiy during neuroleptanalgesia induced with dehydrobenzperidol and phenoperidine. Br J Anaesth 40: 143–144
2. Bickford RG, Faulconer A, Sem-Jacobson CH, Petersen MG, Dodge HW, Schnugg FJ (1953) Some effects of barbiturate anesthesia on depth electrogram. Proc Staff Meet Mayo Clin 28: 162–165
3. Boyan CP, Howland WS, Belloille JW (1956) Electroencephalographic pattern during steroid narcosis. Anesth Analg (Paris) 35:415–417
4. Brazier MAB, Fineginger JE (1945) Action of barbiturates on cerebral cortex: Electroencephalographic studies. Arch Neurol Psychiat 53:51–58
5. Bushart W, Rittmeyer P (1966) EEG-Kontrollen vor, während und nach Neuroleptanalgesie. In: Henschel WF (ed) Die Neuroleptanalgesie. Springer, Berlin Heidelberg New York. p 44–44
6. De Castro J, Mundeleer P (1959) Anesthesia without sleep: Neuroleptanalgesia Acta Chir Belg 58:689–693
7. De Castro J, Mundeleer P (1962) Die Neuroleptanalgesie, Auswahl der Präparate, Bedeutung der Analgesie und der Neurolepsie. Anaesthesist 11: 10–17
8. Chen G, Westen JK (1980) The analgesic and anesthetic effects of l-(-1 phenylcyclohexyl) piperidine-HCl on the monkey. Anesth Analg 39: 132–137
9. Corssen G (1965) Pharmakologische und erste klinische Erfahrungen mit dem Kurznarkotikum CI-581. Tagung Europ. Anästhesiegesellschaft: conference, Zürich
10. Corssen G (1966) Clinical use of CI-581. 2nd European Congr. Anaesthesiol., Copenhagen, 8–13. Aug. 1966
11. Corssen G, Domino EF (1966) Dissociative anesthesia: further pharmacologic studies and first clinical experiments with the phencyclidine derivate CI-581. Anesth Analg 45:29–40
12. Corssen G, Domino EF, Bree RL (1969) Electroencephalographic effects of ketamine anesthesia in children. Anesth Analg 48: 141–147
13. Davis B, Dundee JW, Forrester AC, Johnson SE, Lewis AAG, Simpson BRJ, Sutton JA (1972) Steroid anaesthesia. Postgrad Med 48:2 (suppl): 1–137
14. Doenicke A, Gabanyi HL, Schürk-Beelich M (1974) Kreislaufverhalten und Myokardfunktion nach drei kurzwirkenden i.v. Narkotika (Etomidate, Propanidid, Methohexidal). Anaesthesist 23: 108–115
15. Doenicke A, Kugler J, Penzel G, Laub M, Kalmar L, Killian I, Bezecny H (1973) Hirnfunktion und Toleranzbreite nach Etomidate, einem neuen barbitursäurefreien i.v. applizierbaren Hypnotikum. Anaesthesist 23:357–366
16. Doenicke A, Kugler J, Schellenberger A, Gürthner T (1966) The use of electroencephalography to measure recovery time after intravenous anaesthesia. Br J Anaesth 38:580–590
17. Faber B, Hajuseck I (1973) Propanidid (Epontol®) – A new Activation Method in Electroencephalography. Epilepsia 14:359–362
18. Finesinger JE, Brazier MAB, Tucci JH, Miles HHW (1947) Study of levels of consciousness based on electroencephalographic data in pentotal anesthesia. Trans Am Neurol Assoc 72: 183–185
19. Foldes FF (1973) The preventior of the psychomimetic effects of ketamine. In: Gemperle M, Kreuscher H, Langrehr D (eds) Ketamin. Springer, Berlin Heidelberg New York, pp 251–254 (Anästhesiologie und Wiederbelebung, vol 62)
20. Gibbs FA, Gibbs EL, Lennox WG (1937) Effect on electroencephalogram of certain drugs which influence nervous activity. Arch Intern Med 60: 154–166
21. Grabow L, Pyhel N (1980) Veränderungen der hirnelektrischen Aktivität unter dem Einfluß der allgemeinen Anästhesie. Anaesthesist 29:366–369

22. Gubernatis G (1981) EEG-Befunde im postoperativen Zeitraum unter Berücksichtigung der vorangegangenen Narkoseart. Dissertation Med. Hochschule Hannover
23. Hatano S, Nishiwada M, Matsumura M (1978) Ketamine – Diazepam anaesthesia for abdominal surgery. A Review of 1000 cases with continuous "Micro-Mini" Drip administration technique. Anaesthesist 27:172–182
24. Henschel WF (1966) Die Neuroleptanalgesie. Allg Med Landarzt 42:1614–1623
25. Herrschaft H, Schmidt H (1973) Das Verhalten der globalen und regionalen Hirndurchblutung unter dem Einfluß von Propanidid, Ketamine und Thiopental-Natrium. Anaesthesist 22:486–495
26. Ingvar DH, Cronquist S, Eckberg R, Risberg J, Hoedt-Rasmussen K (1965) Normal values of regional cerebral blood flow in man including flow and weight estimates of grey matter. Acta Neurol Scand 41 (Suppl.) 14:72–78
27. Iwatsucki K (1973) Modified neuroleptanalgesia using droperidol and pentazocine. Int Anesthesiol Clin 11:79–91
28. Janssen P (1962) Colloquium über NLA. Düsseldorf 1961. Anaesthesist 11:1–7
29. Kopferer JM (1973 Kombination von Ketamin mit Valium und DHB. In: Gemperle M, Kreuscher H, Langrehr D (eds) Ketamin. Springer, Berlin Heidelberg New York, p 258 (Anästhesiologie und Wiederbelebung vol 69)
30. Kiersey DK, Bickford RG, Faulconer A (1951) Electroencephalographic patterns produced by thiopental sodium during surgical operations. Description and classification. Br J Anaesth 23:141–152
31. Knoche E, Tranto E, Dick W (1978) Möglichkeiten der medikamentösen Beeinflussung von unerwünschten Nebenwirkungen und Aufwachreaktionen nach Ketamin-Anästhesie. Anaesthesist 27:302–308
32. Kortilla K, Linolla M (1974) Skills related to driving after intravenous diazepam, flunitrazepam or droperidol. Br J Anaesth 46:961–969
33. Kreuscher H (1965) Die cerebrale Durchblutung und Sauerstoffaufnahme beim Hund unter Neuroleptanalgesie. Thesis, University of Mainz
34. Kreuscher H (1977) Erfahrungen mit der Tranquanalgesie. In: Rügheimer E (ed) Erlangen Anästhesie Seminare I:46. Medizin Media Analyse. Henke, Bubenreuth
35. Kreuscher H, Hübner J (1976) Tranquanalgesie. Ketanest-Symposion, Alfeld/Leine West-Germany 1976. Parke-Davis Literaturdienst, p 21–25
36. Kubicki S (1966) EEG-Veränderungen durch Neuroleptanalgesie. In: Gemperle H (ed) Neuroleptanalgesie. Springer, Berlin Heidelberg New York (Anästhesiologie und Wiederbelebung vol 18)
37. Kubicki S, Just O, Götze W (1958) Die Steroid-Narkose und ihr elektroenzephalographisches Bild. Anaesthesist 7:39–44
38. Kubicki S, Stölzel R, Otten I, Haas J (1970) Exzitatorische und inhibitorische Phänomene am Zentralnervensystem verursacht durch Fentanyl. In: Henschel WF (ed) Neue klinische Aspekte der Neuroleptanalgesie. Schattauer, Stuttgart p 21
39. Kubicki S, Zadek P (1966) EEG-Veränderungen durch Neuroleptanalgesie. In: Henschel WF (ed) Die Neuroleptanalgesie. Springer, Berlin Heidelberg New York, p 44–49
40. Kugler A, Doenicke A, Laub H, Kleinert H (1969) Elektroenzephalographische Untersuchungen bei Ketamine und Methohexital. In: Kreuscher H (ed) Ketamine. Springer, Berlin Heidelberg New York, p 101–109
41. Kugler J, Doenicke A, Laub M (1977) The EEG after Etomidate. Anaesthesiol Wiederbeleb 106:31–47
42. Lehmann KA, Daub D (1979) Versuch einer Dosierungsoptimierung von Fentanyl in der Neuroleptanalgesie. Prakt Anaesth 14:293–302
43. Lundy JS (1935) Intravenous anaesthesia: Preliminary report of the use of two new thiobarbiturates. Proc Staff Meet Mayo Clin 10:536–543
44. Mark LG, Brand L, Dayton P, Blaber P, Papper EM (1960) Behaviour of methohexital in man. Fed Proc 19:177
45. Martin JT, Faulconer A, Bickford RG (1959) Electroencephalography in anesthesiology. Anesthesiology 20:359–376
46. Muhlmann-Weill M, Mangene F, Ganthier-Lafage JP (1977) Die Bedeutung der Kombinationsnarkose mit Ketamin und Diazepam. Anesth Analg Reanim 29:355–363
47. Nilsson E, Ingvar DH (1967) EEG-findings in neuroleptanalgesia. Acta Anaesthesiol Scand 11:121–127

48. Oberwetter WD, Hoffmann H, Degen R, Opitz A, Degen HE (1980) Neuroleptanalgesie bei Patienten mit zerebralem Krampfleiden. In: Opitz A, Degen R (eds) Anästhesie bei zerebralem Krampfleiden und Intensivtherapie des Status epilepticus. Perimed, Erlangen, p 119–122
49. Orosz E, Toth S, Juhasz J (1972) Der Einfluß der Neuroleptanalgesie auf die elektrische Aktivität des Gehirns. EEG EMG 2:76–84
50. Osswald PM, Hartung HJ (1980) Klinische Pharmakologie der intravenösen Einleitungsnarkotika. Anaesthesiol Intensivmed 1:9–16
51. Paeschke D (1978) Naloxon. Prakt Anaesth 13:127–134
52. Patschke D, Brückner JB, Gethmann JW, Stoijiljkovic D, Tarnow J, Weymar A (1975) Einfluß von Methohexitol und Thiopental auf die Coronarperfusion und die Sauerstoffversorgung des Herzens. In: Bergmann H, Blauhut B (eds) Respiration, Zirkulation, Herzchirurgie. Springer, Berlin Heidelberg New York, pp 74–78 (Anästhesiologie und Wiederbelebung vol 93)
53. Pichlmayr I, Drost R, Soga D, Beer R (1970) Über das Verhalten der Hirndurchblutung des Hundes unter Narkose mit Propanidid und Methohexital-Natrium. Anaesthesist 19:144–146
54. Price HL (1960) A dynamic concept of the distribution of thiopental in the human body. Anesthesiology 21:40–45
55. Price HL, Kornack PJ, Safar JN, Conner EH, Price ML (1960) The uptake of thiopental by body tissues and its relation to the duration of narcosis. Clin Pharmacol Ther 1:16–22
56. Schwartz MS, Virden S, Scott DF (1974) Effects of Ketamine on the electroencephalograph. Anaesthesia 29:135–140
57. Schomburg ED, Baiker-Heberlein M (1975) Neurophysiologische Untersuchungen zur zentralnervösen Wirkung des Hypnotikums Etomidate an der Katze. Anaesthesist 24:269–272
58. Schwilden H, Stoeckel H (1980) Untersuchungen über verschiedene EEG-Parameter als Indikatoren des Narkosezustandes. Anaesthesiol Intensivther Notfallmed 15:279–286
59. Scott JW, Gordon RA (1956) EEG during anaesthesia with ketosteroid (Viadure) Electroencephalog. Clin Neurophysiol 8:525–533
60. Scott DF, Virden S (1972) Comparison of the effect of Althesin with other induction agents on electroencephalographic patterns. Postgrad Med J 48:93–96
61. Shimazono Y, Ohuma T, Fukuda T, Hirai T, Yamamasu E (1953) Electroencephalographic study of barbiturate anesthesia. Electroencephalogr Clin Neurophysiol 5:525–532
62. Stoelting VK (1957) The use of a new intravenous oxygen barbiturate 25 398 for intravenous anesthesia. Anesth Analg Curr Res 36:49–51
63. Takeshita H, Miyauchi F, Ishikawa T (1973) Canine cerebral oxygen consumption studies related to electroencephalogram during propanidid anesthesia. Acta anaesthesiol Scand 17:227–237
64. Weese H, Scharpf W (1932) Evipan, ein neuartiges Einschlafmittel. Dtsch Med Wochenschr 58:1205–1207

V. Muscle Relaxants

Today, muscle relaxants are the most important adjunct in anesthesiology. Their purpose is to maintain muscular relaxation over a long period at a predetermined level of anesthesia appropriate for surgery. Consequently, both the patient's condition and perioperative requirements have to be considered. In intensive care, numerous therapeutic procedures also have to be taken into account. Broadly, two groups of relaxants with different mechanisms of action are available for clinical use – depolarizing and nondepolarizing substances (LEE and ATKINSON 1978, BENZER et al. 1982).

Pharmacological Mode of Action and Clinical Application. The peripheral, and usually exclusive, site of action of muscle relaxants is the neuromuscular end-plate. Nerve impulses are transmitted onto the muscle tissue via acetylcholine liberation and specific reactions with receptors. Depolarizing substances (main representative – succinylcholine and decamethonium), by means of loose receptor binding, cause a depolarization of the end-plate, which is transmitted to the muscle fibers. As a result of the (electrically negative) depolarization, the muscle is refractory to stimuli. The relaxant substances are hydrolytically metabolized by serum cholinesterases. Specific antidotes, e.g., esterases, are available but are not as yet suitable for common clinical practice.

Nondepolarizing muscle muscle relaxants (main representatives: – tubocurarine, alcuronium, pancuronium) inhibit the effects of acetylcholine at the neuromuscular end-plate by reacting with the receptor protein. This effect is reversed by biological inactivation and competition with acetylcholine at the receptor site. Cholinesterase inhibitors have antagonizing effects. Accumulation of acetylcholine causes a competitive removal of the relaxants from the receptors. Muscle relaxants lead to a muscular atony in a dose-dependent matter, which can be employed clinically. Areas are affected by atony in the following order – head, jaw, neck, extremities, trunk, respiratory muscles. Recently developed muscle relaxants involve only mild cardiovascular, vegetative, and humoral side effects. Central actions do not normally occur because an intakt blood-brain barrier is impermeable to these substances [1, 3–5, 8].

EEG Findings

Our Findings (500 patients 20–80 years of age) (Fig. 1). During encephalographic monitoring of anesthesia, 500 patients were given a short-acting depolarizing muscle relaxant (succinylcholine) for intubation. Intraoperatively, nondepolarizing muscle relaxants with a longer effective period, alcuronium, and pancuronium, were used as needed. The EEG showed both groups to suffer no visible effects on cerebral function. The occasional impression of an encephalographic intensification of anesthesia after succinylcholine administration was due to the disappearance of muscular artefacts in the EEG as a result of relaxation.

Review of Other Findings. The above observations have been confirmed by other study groups. They found that in the absence of hypotension, hypothermia, hypoxia, or hyper- and hypocapnia, relaxants or even relaxant antidotes did not have any effects on the EEG (BRECHNER et al. 1962 [2, 6, 9]). On the other hand, after application of depolarizing muscle relaxants with a simultaneous increase in pulse rate and blood pressure, arousal reactions in the EEG could be seen [7].

After application of nondepolarizing muscle relaxants, changes in alpha reactivity subsequent to visual stimulation were observed [2]. In animal experiments, VON WILD [10] found that with an intact CNS and blood-brain

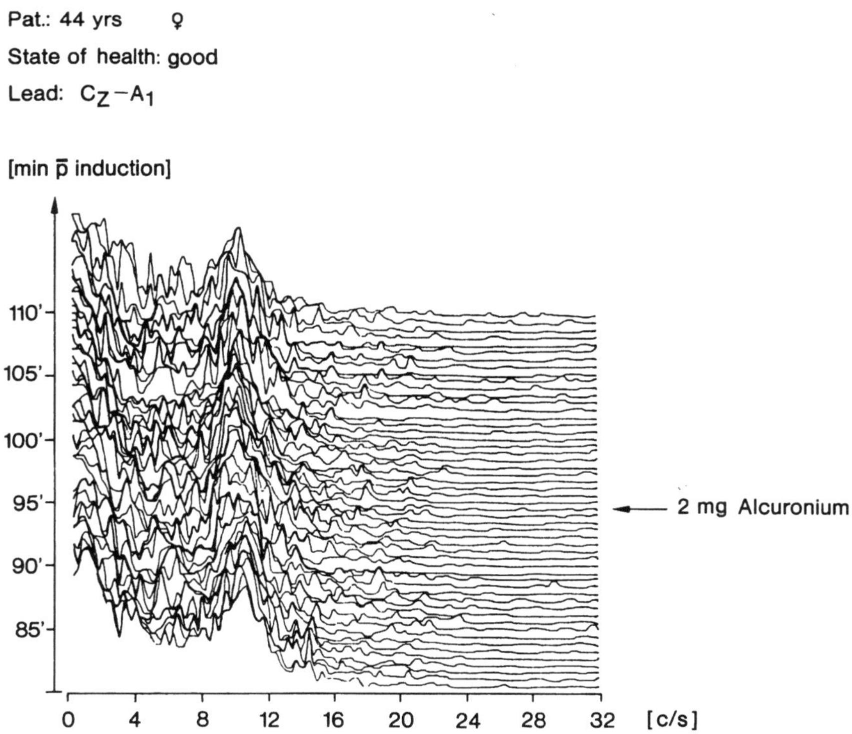

Fig. 1. EEG spectral analysis of neuroleptanalgesia in steady state; 2 mg i.v. alcuronium do not lead to visible changes in the EEG background activity.
Leads: C_Z-A_1

barrier, direct and indirect encephalographic relaxant effects were absent and that the inhibition of the hypothalamus was due to a decrease in afferent stimuli. On the other hand, after cerebral damage and with an impaired blood-brain barrier, cerebral changes could be observed to a significant extent.

Cerebral reactions occur more frequently with depolarizing than with nondepolarizing substances. With a simultaneous increase in the peripheral circulatory parameters, approximately 20 s after intravenous relaxation, arousal reactions with desynchronization (increase in frequency, decrease in amplitude) become visible in the cortical EEG as does synchronization (decrease in frequency, increase in amplitude) in the hippocampal EEG. This occurs for 40–60 s. Paroxysms of electric cerebral activity that might already be present, as manifestation of an increased excitability, are not affected, and are not prolonged even in the case of an impaired blood-brain barrier.

Clinical Evaluation. Muscle relaxants are used according to their clinical affect. Dosage is determined by the physical data of the patient. It is affected by physiological changes, the depth of anesthesia, and synergistic or potentiating effects of the main anesthetic. Immoderate application can lead to excessive relaxation, with the possibility of postoperative respiratory insufficiency.

Interpretation of EEG Changes and Clinical Findings. Muscle relaxants can be used as required in the cerebrally healthy patient. In the case of disturbances in cerebral function, central effects in terms of a brief arousal reaction may be expected after the application of depolarizing relaxants and occasionally after the injection of nondepolarizing muscle relaxants. There are no cerebral electric signs of seizure. Negative effects on cerebral function do not normally occur. Muscle relaxants can, therefore, also be used without hesitation for neurosurgical procedures or in intensive care in patients with an impaired blood-brain barrier (e.g., in septicemia). The same applies to infants, whose blood-brain barrier is not fully formed until the age of 2 years.

References

A. Textbooks and Summaries

Benzer H, Frey R, Hügin W, Mayrhofer O (1982) Lehrbuch der Anästhesiologie, Reanimation und Intensivtherapie. Springer, Berlin Heidelberg New York

Brechner VL, Walter RD, Dillon JB (1962) Practical electroencephalography for the anesthesiologist. Thomas, Springfield

Lee JA, Atkinson RS (1978) Synopsis der Anästhesie. Fischer, Stuttgart

B. Papers

1. Cohen EN, Hood N, Golling R (1968) Use of wholebody autoradiography for determination of uptake and distribution of labelled muscle relaxants in the rat. Anesthesiology 29:987–993
2. Davis HS, Dillon WH, Collins WF, Randt CT (1958) Effect of anesthetic agents on evoked central nervous system responses: muscle relaxants and volatile agents. Anesthesiology 19:441–449
3. Feldmann SA (1973) Muscle relaxants. Saunders, London 1–150
4. Foldes FF (1972) The rationed use of neuromuscular blocking agents: the role of pancuronium. Drugs 4:153–162
5. Grob D (1967) Neuromuscular blocking drugs. Effect on central nervous system. In: Root WS, Hoffmann FG (eds) Physiological pharmacology. Academic Press, New York p 449–454
6. Kiersey DK, Bickford RG, Faulconer A jr (1951) Electroencephalographic patterns produced by thiopental sodium during surgical operations. Description and classification. Br J Anaesth 23:141–152
7. Mori K, Iwabuchi K, Fujita M (1973) The effect of depolarizing muscle relaxants on the electroencephalogram and the circulation during halothane anaesthesia in man. Br J Anaesth 45:604–610
8. Smith SM, Brown HO, Toman JEP, Goodman LS (1947) Lack of cerebral effects of d-tubocurarine. Anesthesiology 8:1–14
9. Shea LT, Davidson GM, Davis J (1954) Electroencephalographic studies of curarized patients. Med J Aust 2:656–659
10. von Wild K (1980) Tierexperimentelle Untersuchungen über die Beeinflussung des EEG durch Muskelrelaxantien am Beispiel von Pancuronium und Succinylbischolin. In: Opitz A, Degen R (ed) Anästhesie bei zerebralen Krampfanfällen und Intensivtherapie des Status epilepticus. Perimed, Erlangen p. 45–55

VI. Analgetics in the Postoperative Phase

CONTENTS

Short-term or long-term management of pain is frequently required in the immediate postoperative period as well as in the later stages. It is particularly essential for the general recovery of the patient and for normalization of the vital functions after extensive surgery. Individual pain perception is not only affected by the type and degree of pain stimulus and its spinal transmission, but also by identification of pain in the limbic system, pain localization in the postcentral cortex, and the processing of pain in the frontal cerebrum (Fig. 1) [15]. Since the extraction of the opium alkaloid morphine by SERTÜRNER 1805–1817 [27], centrally acting analgetics have been applied intravenously, intramuscularly, and subcutaneously for postoperative pain management; they are gradually reduced in strength up to usually the 3rd day depending, of course, on the extent of surgery. In additional, the newly developed psychomimetic drugs may be used therapeutically for their potentiating, tranquilizing, or anxiolytic properties. The postoperative mitigation of pain is possible, as has recently been discovered, by repeated doses of very small amounts of morphine dissolved in a certain volume of a normal saline solution and administered by means of a peridural indwelling catheter; this produces a reversible, "dissociated analgesia." This kind of treatment is based on the discovery of specific opium receptors in the gelatinous substance of the posterior horn of the spinal cord, and brings about a morphine-induced reversible interruption of pain-conducting c-fibre afferences [3, 28, 32, 35–39].

The major site of action of central analgetics is the mesencephalic area, where pain transmission to the limbic system is suppressed or even discontinued by depression of the neuronal activity. The simultaneous suppression of the ascending reticular system leads to a decrease in vigilance.

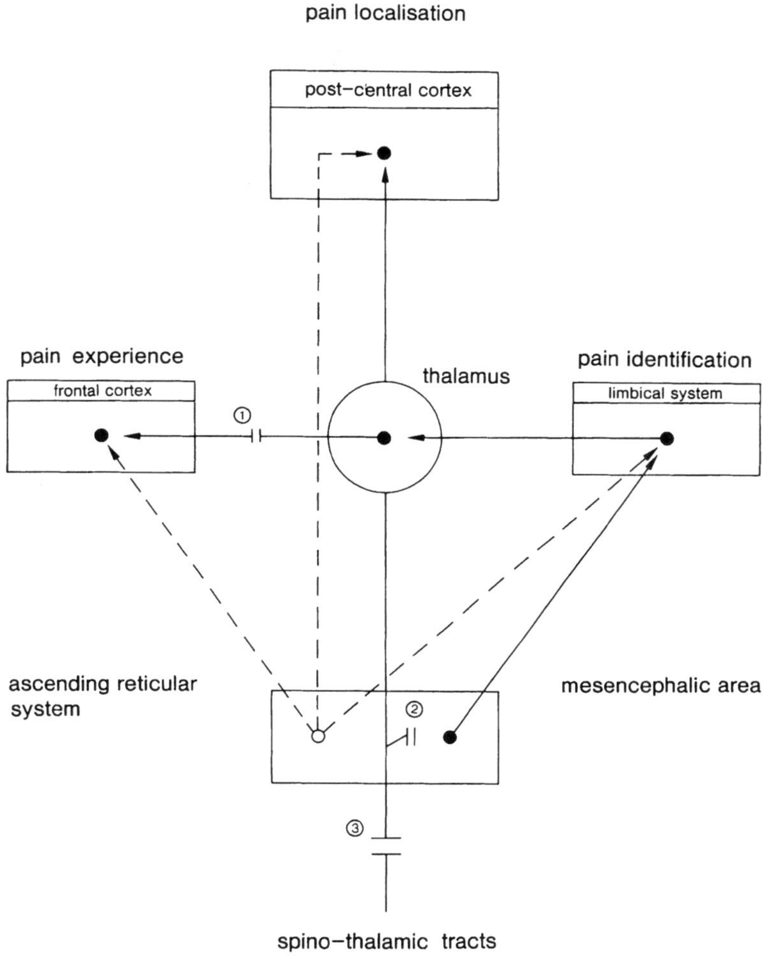

Fig. 1. Pain analysis in the central nervous system. The localization of pain depends on the intact transfer of the afferences of the spinal-thalamic tract via the thalamus to the areas of the postcentral cortex. The identification of the pain results through switching of the afferences in the mesencephalic area and a transfer to the limbic system. From there, after further switching in the thalamus and conduction to the frontal cortex, the pain experience is formed. Mesencephalic transfer to the ascending reticular system (ARS) leads to an unspecific increase in vigilance at the onset of a painful event.
1, Severence of the thalamo-frontal pathways (leukotomy) inhibits the experience of pain, but does not exclude the localization and identification of pain.
2, Blockage of the mesencephalic switching to the limbic system (target of the analgetics) results in loss of the pain identification only. It does not, however, prevent the pain experience and the localization.
3, When severing the spinal thalamic tract, all three pain-processing systems are blocked. (Modified from KUBICKI [15])

Pontine and diencephalic functions are also affected to a lesser degree. The specific analgetic potency of central analgetics has been studied widely, and each drug has been compared and labeled according to the analgetic effect of morphine. The analgetic effect of the substances cannot be determined from the EEG; one can, however, assess any centrally suppressing effects that analgetics might be exerting. These effects can be used to advantage postoperatively in some cases. The effects of common analgetics will be presented as examples.

Morphine (Morphium hydrochloricum)

Pharmacological Mode of Action and Clinical Application. The strong specific analgetic efficiency of morphine with its simultaneous euphoristic properties has been the reason for its almost unlimited application. Even today, therapy of severe, distressing pain is carried out with morphine. The mitigation of pain perception takes place in the central nervous system without considerably interfering with other sensory stimuli. The cerebral cortex and diencephalic structures, especially the thalamus and mesencephalic areas, are assumed to be the sites of action. Analgesia is accompanied by varying degrees of hypnotic effects caused by a general central depression. With a high dosage, this may extend to a deep sleep or coma. The emotional state is frequently elevated, probably as a result of the eradication of unpleasant sensations; this effect, however, may be absent, or may even be reversed. In long-term use as may be necessary, for example, in intensive care, tachyphylaxia and habituation with signs of addiction occur. Central nervous suppression includes the respiratory center. The major morphine effect, which is also the most dangerous, is a sustained suppression of ventilation, even with common therapeutic use. Apart from central inhibitory properties, centrally exciting effects occur, especially on the vomiting center and on cholinergic parts of the occulomotor nerve. This leads to nausea and occasionally to vomiting and miosis. Extreme excitation through central stimulation by morphine rarely occurs in man. The spectrum of peripheral effects includes increased tone of the gastrointestinal tract and vesical sphincter, release of antidiuretine, and an increase in sympathetic drive. Morphine does not suppress circulation. The therapeutic single dose is 0.1–0.2 mg/kg.

EEG Findings

Our Findings (20 patients, 20–50 years of age, dose 5 mg or 10 mg morphine i.v. in each of ten patients (Table 1; Figs. 2, 3) [17]). No change is seen after intravenous morphine application in about 30% of the EEG findings. With low doses of approximately 0,08 mg/kg, there is a slight re-

Table 1. EEG changes after intravenous application of morphine ($n = 20$)

Frequency range	EEG power change after drug application[a]	Incidence (%)	Degree of change	Morphine dose
Alpha	↓↑ a)	50	5–10 μV	5 mg = approx 0.08 mg/kg
	↓ b)	20		
	↑ c)	20		
	0 d)	20		
Beta	0	0	0	
Delta	(↑)	33	approx. 5 μV	
Theta	(↑)			
Alpha	↓↑ a)	60	5–20 μV	10 mg = approx 0.14 mg/kg
	↓ b)	10		
	↑ c)	10		
	0 d)	10		
Beta	↑	10	approx. 20 μV	
Delta	(↑)	60	approx. 10 μV	
Theta	(↑)			

[a] ↑ increase, ↓ decrease, (↑) light increase, 0 absence

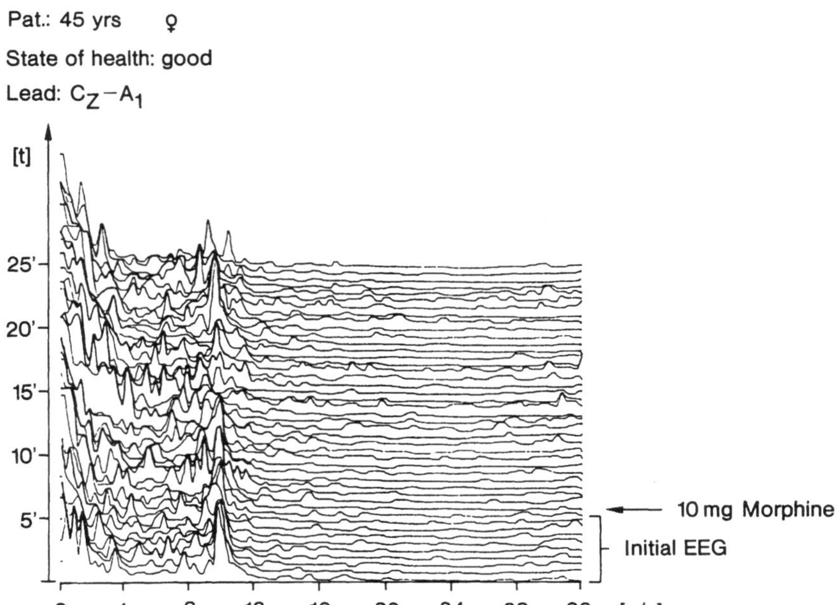

Fig. 2. EEG spectral analysis prior to and after i.v. injection of 10 mg morphine. The initial alpha EEG shows an average frequency of 10 Hz with normal voltage. After injection, the electric power of the alpha band tends to show alternations. During the reduction in alpha power, an increase in activity in the delta/theta band was seen. The clinical sedative effect was very good (+++).
Recording conditions: TC, 0.3 s; filter, 70 Hz; calibration, 50 μV = 7 mm; paper speed, 30 mm/s; FFT, 30-s periods

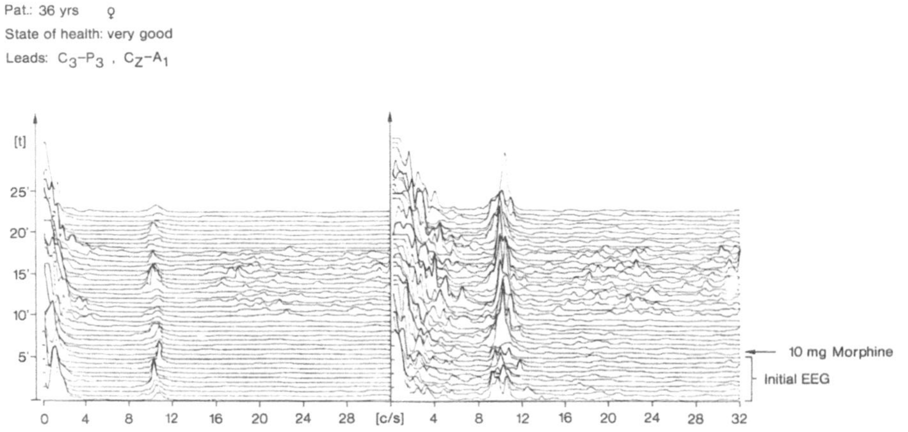

Fig. 3. EEG spectral analysis prior to and after i.v. injection of 10 mg morphine. The initial alpha EEG, in both tracks, has an average frequency of 10–11 Hz. After injection, varying voltages of the alpha band are observed with corresponding increased activity in the delta/theta band. As manifestation of the excitatory component of the morphine, there is a marked increase of activity in the beta range 5–10 min after injection. No clinical signs of sedation. Recording conditions: TC, 0.3 s; filter, 70 Hz; calibration, 50 μV = 7 mm; paper speed, 30 mm/s; FFT, 30-s periods

duction in the voltage (5–10 μV) of the initial frequency. This does not occur with a higher dose of about 0.14 mg/kg. Generally, in 70% of the cases, increases or decreases in power are observed in the range of the initial frequency; they are between 8 and 20 μV and are dose-dependent. An increase in the lower frequencies (delta/theta) is on the whole less marked; it may occur intermittently in approximately 30–70% of patients, also in a dose-dependent manner. In the group receiving the higher dose, the formation of clear beta activity was observed in one patient, indicative of the stimulating properties of morphine (Fig. 3).

Review of Other Findings. EEG findings in psychiatric patients (previous alcoholics) with a subcutaneous application of 20 mg morphine indicate no change or only occasional sleep patterns. The level of consciousness of subjects affected by morphine does not always parallel the registered EEG changes [1]. Intravenous morphine administration frequently enhances preexisting abnormal EEG findings [24]. After a slow i.v. application of 30 mg morphine, WIKLER [33] observed clinical signs of erythema. It was accompanied by pleasant feeling, excitement, and a notable state of alertness lasting for 2–5 min. EEG changes were not seen. This initial phase is followed by sustained changes in consciousness, varying both in direction and degree, with simultaneous power changes in the EEG. In the majority of cases, relaxation and euphoria are accompanied by somnolence, rarely by increased alertness. Here the EEG shows synchronization. Occasionally, clinical signs of depression and fear are recorded with simultaneous desynchronization in the EEG. Cessation of clinical responsiveness is also observed in the EEG. Wikler [33] claims that EEG synchronization and desynchronization reflect activity changes in neuronal systems, which are loosely related to behavioral changes. The latter also seem to depend on cortical homoeostasis. Desynchronization occurs when there is an excessive increase in activity in these structures and can be interpreted as an inhibitory effect on conscious-promoting systems; synchronization has the reverse effect.

Clinical Evaluation. Morphine does not alter a subject's sense of well-being or consciousness in approximately 30% of patients. In the remainder, mild calming usually occurs, only occasionally, does heavier sedation take place. Euphoria was not observed. The peripheral circulation was little affected; it did not change significantly in either group. Side effects were frequent, especially with the lower doses of morphine (60% with 0,08 mg/kg, 40% with 0.15 mg/kg); they include nausea, dizziness, fear, restlessness, and occasional circulatory reactions with outbreaks of cold sweat.

Interpretation of EEG Changes and Clinical Findings. The incidence and degree of frequencies in the delta/theta band correspond to the sedative clinical effects. However, a strict correlation between a patient's subjective feeling and the observed power changes of the basic frequency is not evident. Thus, with morphine, though the occurrence and degree of sedation may be determined from the EEG, the power changes in the initial frequency must be regarded as changes in vigilance, spontaneously induced by morphine. They mostly coincide with unpleasant side effects and in an individual case do not correspond with clinically determinable changes in consciousness or well-being.

Pethidine (Dolantin)

Pharmacological Mode of Action and Clinical Application. Pethidine was first introduced into clinical therapy in 1939 as the first synthetic drug with morphinelike properties. Even today, it still belongs to the pain-relieving drugs in very common use. Its central and peripheral signs of action are largely similar to those of morphine; analgetic efficiency and other effective components, including side effects, are less pronounced and of shorter duration. With an equipotent analgetic dosage, pethidine has no significant advantages over morphine. For a 10-mg dose of morphine, five to ten times this amount would be required of pethidine to achieve the same analgetic effect (Benzer et al. 1982; Kuschinsky and Lüllmann 1974).

Pat.: 77 yrs ♀

State of health: good

Lead: C_3–P_3

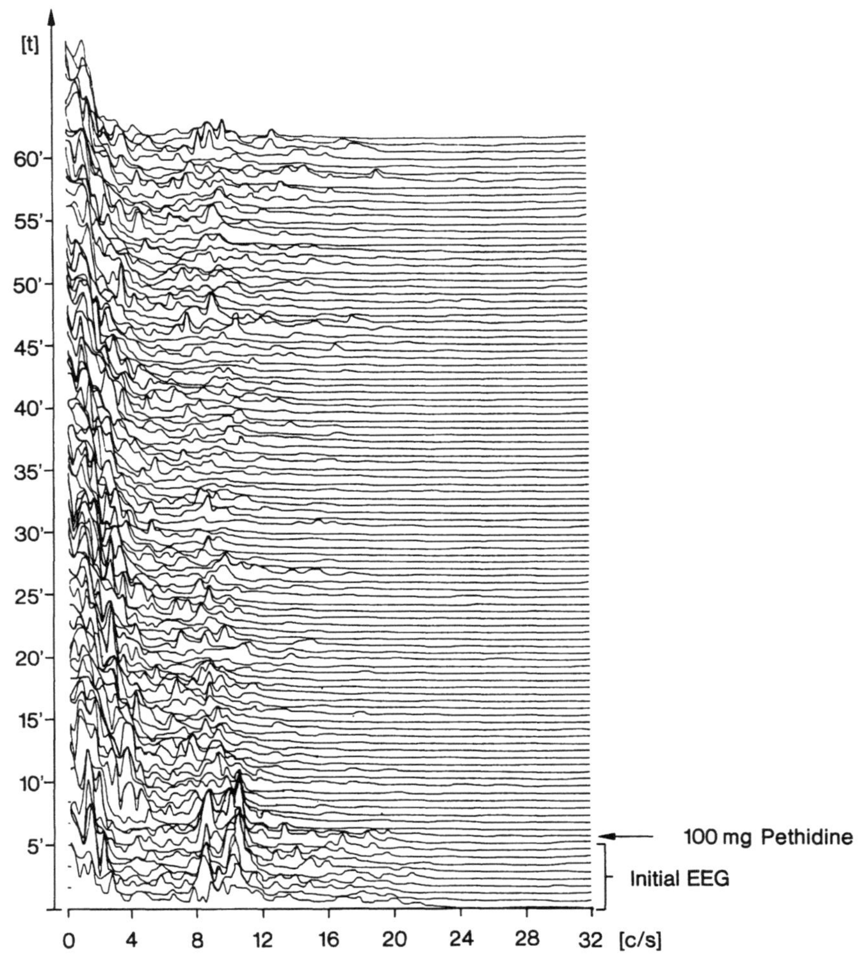

Fig. 4. EEG spectral analysis prior to and after injection of 100 mg pethidine. After injection, the preexisting alpha EEG is reduced and there is a power increase in the range of 2–4 Hz. An increase of voltage in the alpha band is again observed after approximately 40 min. Excellent clinical sedative effect (+ + +).
Recording conditions: TC, 0.3 s; filter, 70 Hz; calibration, 50 μV = 7 mm; paper speed, 30 mm/s; FFT, 30-s periods

Table 2. EEG changes after intravenous injection of pethidine ($n = 16$)

Frequency range	EEG power change after drug application[a]	Incidence (%)	Degree of change	Morphine dose
Alpha	↓ or ↓↑	100	15–70 μV	
Beta	↓ upper frequency range (with irregular initial EEG)	100	10–20 μV	100 mg = approx 1.6 mg/kg
Delta	↑	69	15–40 μV	
Theta	↑			

EEG Findings

Our Findings. The EEG findings and clinical changes under intravenous administration of 100 mg pethidine are described in Chap. BII [20] (Table 2, Fig. 4). There is a decrease in power in the dominant basic frequency, which is especially pronounced in young patients during the first 10–15 min, and there are occasional delta and theta waves. These initial EEG changes do not disappear completely even over a longer observation period of up to 50 min. The shift of the dominant frequency peak, by 0.5–1 Hz, into the lower frequency range occurs and persists. The ability to assess the psychic, relaxing, and sedative effects of pethidine from the EEG pattern is good. As with other analgetics, the existence and degree of analgesia cannot be detected in the EEG.

Piritramide (Dipidolor)

Pharmacological Mode of Action and Clinical Application. Piritramide is also a synthetic opioid. In 1961, it was introduced by Janssen as a potent substance suitable for the treatment of pain [12]. Its specific analgetic effect leads to good or very good clinical results in 78% of cases. Though weaker, it acts for a longer time than morphine. Undesired central and peripheral side effects, such as respiratory depression, vomiting, dizziness, and decrease in blood pressure, are mild or absent. Thus, piritramide is the analgetic of choice today for various indications, especially in the circulatory and cardially impaired patient. A 15-mg dose of piritramide is regarded as equipotent to 10 mg morphine [11, 14, 19, 23, 29, 30].

EEG Findings

Our Findings (16 patients – eight patients 20–50 years of age, eight patients over 70 years, dose 15 mg = 2 ml piritramide base, given i.v. within 2 min)

Table 3. EEG changes after intravenous injection of piritramide ($n = 16$)

Frequency range	EEG power change after drug application[a]	Incidence (%)	Degree of change	Piritramide dose
Alpha	↓↑ a)	44	15–20 μV	
	↓ b)	19		
	↓ c)	13		
	d) ↓ dominant frequency	75	0.5–2 Hz (more pronounced in older patients)	
Beta	↑↓	With irregular initial EEG, especially in older patients	15–20 μV	15 mg = approx 0.23 mg/kg
Delta	↑	94	10–30 μV	
Theta	↑	Especially in older patients	10 μV	

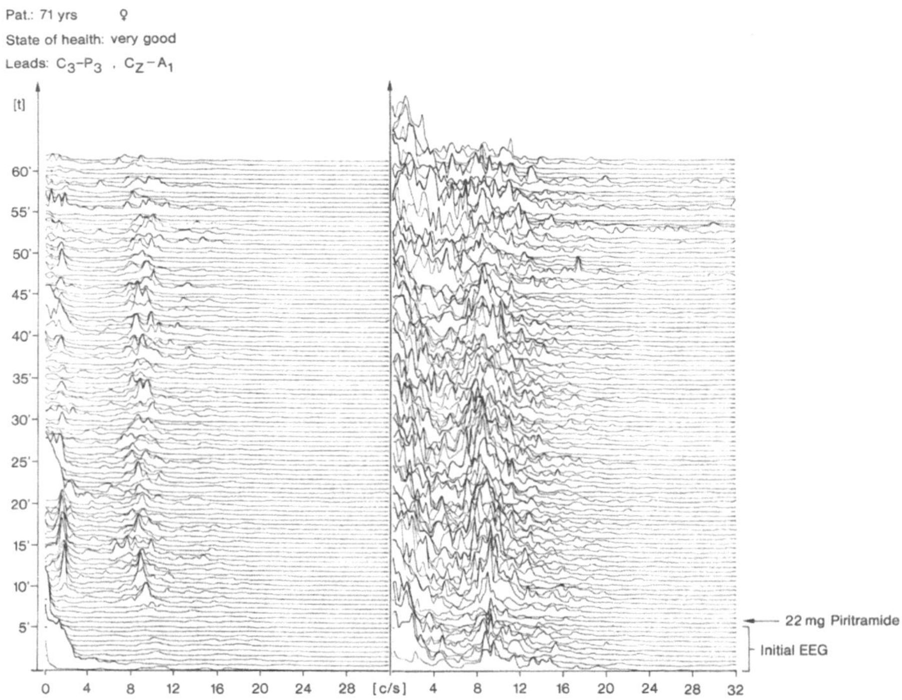

Fig. 5. EEG spectral analysis prior to and after injection of 22 mg piritramide. There is a marked increase in activity of the preexisting alpha rhythm with a dominant frequency of 9 Hz. Later, variations in the alpha power and intermittent delta activity occur, which is typical for piritramide. Clinically very good sedation (+ + +).
Recording conditions: TC, 0.3 s; filter, 70 Hz; calibration, 50 μV = 7 mm; paper speed, 30 mm/s; FFT, 30-s periods

(Table 3, Fig. 5). The relevant EEG findings with piritramide are an intermittent increased or decreased dominant background frequency, and, more rarely, a decrease or increase in power alone. In both cases, the changes are approximately 15–20 μV. In addition, in the majority of geriatric patients, there is a decrease in frequency of 0.5–2 Hz leading to a shift in the background frequency from the alpha to the theta range. At the same time, in all patients, the energy in the delta range – in older patients also in the theta range between 3.5 and 6 Hz – increases by approximately 10 μV.

Clinical Evaluation. Piritramide gives rise to sensations of relaxation and well-being, combined with a mild or strong degree of sedation and sleep in approximately every second patient. The state of pleasant relaxation is cited equally frequently both by young and geriatric patients. The level of sedation is greater in older patients. The incidence of occasional side effects, such as respiratory depression, perspiration, or vomiting, is similar in both age-groups. On the whole, side effects are significantly rarer (Table 1).

Interpretation of EEG Changes and Clinical Findings. According to reports of ITIL and FINK [10], the decrease in alpha amplitude is correlated with a suppression of emotions, perceptual ability, and ideas; increase in alpha amplitude is correlated with relaxation, a feeling of well-being, and euphoria. With piritramide, as is also evident from the EEG, both states of consciousness occur intermittently. This combination of the two stages of vigilance is found by the patient to be pleasant. However, there is no explanation from the EEG for the unpleasant sensations associated with similar changes when morphine is used. In accordance with the pattern of the background activity, changes in consciousness rarely occur in one direction only. The simultaneous decrease in alpha frequency, which is especially marked in geriatric patients, suggests a decrease in vigilance. The varying, occasionally very pronounced increase in activity between 2 and 4 Hz, reflects the degree of sedation or sleep (between 2 and 6 Hz in older patients). The level of consciousness and the degree of sedation with piritramide can be determined from the EEG. Again, the specifically analgetic effect of the drug is not evident from the EEG.

Pentazocine (Fortral)

Pharmacological Mode of Action and Clinical Application. Pentazocine chemically belongs to the group of morphine antagonists. Despite weak morphine antagonizing properties, it acts centrally as a strong analgetic and has at the same time a slight dysphoretic and sedative effect. With equal analgetic doses, pentazocine has lower side effects on the vomiting center than the other analgetics mentioned above, though the effects on the respiratory center are just as severe. With repeated doses, cumulative depressive effects on the respiration are thought to be absent. Very high doses involve psychomimetic disturbances. Peripheral side effects – especially cardiodepression – are largely absent. The average duration of action is 4 h. In the postoperative treatment of pain, pentazocine administration is specially suitable following combination anesthesia with fentanyl. This is due to the fact that along with its strong analgetic effects, it simultaneously antagonizes the undesired central side effects of fentanyl, especially the depression of the respiratory center [16, 22]. In intravenous administration, 0.5 mg/kg pentazocine is given. A single dose of 30 mg, equivalent to the analgetic effect of 10 mg morphine, is rarely exceeded [13, 18, 34].

EEG Findings

Our Findings (16 patients, eight patients 20–50 years of age, eight patients over 70 years; dose 30 mg = 2 ml Fortral i. v. within 1 min) (Table 4; Fig. 6). EEG changes with pentazocine are dose-dependent and vary according to the individual weight-related application. The common clinical single dose of 30 mg is equal to a weight-related dose of 0.35–0.55 mg/kg in our own patients. Approximately 1 min post injection, a transient or sustained decrease in power of 10–30 μV is observed in the dominant initial frequency; with a dose of under 0.35 mg/kg this shift fails to appear. After a further latency of 3–4 min, an increase in power of between 10 and 20 μV (occasionally up to 70 μV) occurs in the delta (theta) range. The EEG changes are more pronounced in geriatric patients.

Review of Other Findings. Pentazocine is a noneuphoric analgetic with mild sedative properties. The central depression, according to Kubicki et al. [16], is not manifested in a deceleration of frequency, but in a transitory formation of spindles. The degree of sedation is dose-dependent. After a single dose of 30 mg, indifference, somnolence, and, frequently, a light sleep are induced. Occasionally, pleasant dreams are reported. This psychomimetic effect is more frequent and less pleasant when the dose is increased. The centrally mitigating effect of pentazocine is regarded by Kubicki et al. [16] as a mesencephalic influence without diencephalic participation.

Pentazocine was tested in doses of 1.5–3.5 mg/kg, with particular regard to its fentanyl-antagonizing properties. When given in the hypnotic fentanyl phase, the substance antagonizes the fentanyl-induced EEG changes. Furthermore, it leads to excitatory signs sustained for 10 min and, occasionally, to increases in frequency; it leads to paroxysmal patterns when injected rapidly. Subsequent formation of beta spindles is an expression of central depression. The clinical effects on consciousness, observed along with analgesia, and the EEG changes depend on the method of application and the individual dose. Hence, the differences in the reported alterations in the

Table 4. EEG changes after intravenous injection of pentazocine ($n = 16$)

Frequency range	EEG power change after drug application[a]	Incidence (%)	Degree of change	Pentazocine dose
Alpha	a) ↓ sustained b) 0	74 25	10–30 μV	
Beta	0	0	0	30 mg = approx 0.46 mg/kg
Delta	↑	63	10–15 μV up to 70 μV (in older patients)	
Theta	(↑)			

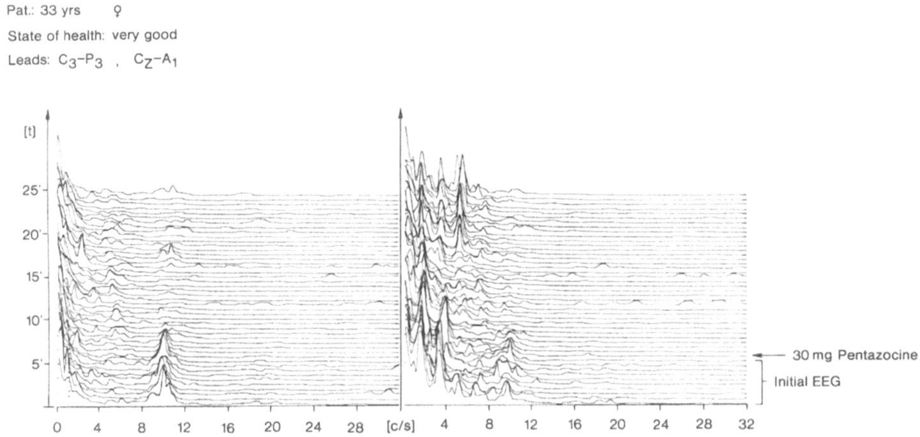

Fig. 6. EEG spectral analysis prior to and after i.v. injection of 30 mg pentazocine. Initial alpha EEG with dominant frequency of 9–10 Hz (the 2–4 Hz activities in the lead C_Z-A_1 are artefacts as a result of movements). After application of pentazocine, the alpha band is suppressed almost completely, and a 5–6 Hz activity established. Clinically mild sedation; side effect palpitations.
Recording conditions: TC, 0.3 s; filter, 70 Hz; calibration, 50 μV = 7 mm; paper speed, 30 mm/s; FFT, 30-s periods

EEG and consciousness can be explained. With a single dose of 30 mg pentazocine, as used in our cases, increases in frequency, formation of spindles, and EEG patterns suggesting an increased tendency for convulsions were not observed.

Clinical Evaluation. Immediately after injection, pentazocine leads to calming and indifference as well as visible signs of sedation, which occurs some minutes later. The clinical effects are more pronounced in the older age-group. Side effects of various degree occur in approximately 25% of patients, equally distributed among the young and geriatric patients. These side effects include dizziness, perspiration, fear, occasionally respiratory depression, and vomiting; they are due to overdose.

Interpretation of EEG Changes and Clinical Findings. Extent and pattern of the EEG changes generally correlate with a subjective change in well-being. The unaffected EEG pattern or the, more frequent, decrease in power in the dominant initial frequency are reflected in a balanced emotional state or calming and indifference, which initially remain constant. The increase in frequency and amplitude, occurring some minutes later in the lower frequency range, leads to subjectively noticeable and objectively visible sedation. Clinical side effects due to individual overdose are not detectable as such in the EEG; the frequency pattern in the alpha and delta bands only shows a strong response to the high dose. The existence and degree of tranquilizing and sedative pentazocine effects can be determined from the EEG; the analgetic effect is not evident in the EEG.

Peridural Application of Analgetics

Pharmacological Mode of Action and Clinical Application. Peridural morphine analgesia is one of the newest methods of pain management. Essential for its clinical efficiency is the existence of specific receptors for pain-relieving drugs in the gelatinous substance of the spinal cord. In peridural analgesia, the required amount of analgetic can be considerably reduced by application near the site of action. Injection of small amounts of morphine dissolved in a larger volume of normal saline solution and applied via an indwelling catheter leads to sustained intensive analgesia [21, 35–39]. The application of peridural anesthesia is clinically indicated in treating pain over longer periods of time in that it decreases the impairment of vigilance and can control analgesia in selective segments. Another important indication is in the therapy of intractable diseases when even larger amounts of systemically applied analgetics no longer lead to pain relief. Here, the precise mechanism of action at the opiate receptors has not yet been established. Approximately 15 min after application of the analgetic, high concentrations of the substance are found in the cerebrospinal fluid with only low plasma levels [8]. The incidence of cerebral side effects, such as nausea, vomiting, and respiratory depression, is given as 32–40% [2, 4–7, 9, 25, 26, 31]. Respiratory depression may occur hours after application of the analgetic. A possible explanation for this is the duration of circulation of cerebrospinal fluid, which takes about 6 h.

EEG Findings

Our Findings (ten patients, dose 5 mg morphine in 10 ml normal saline, applied via a lumbar peridural indwelling catheter, observation period

Table 5. EEG changes after peridural application of morphine ($n = 10$)

Frequency range	EEG power change after drug application[a]	Incidence (%)	Degree of change	Peridural morphine dose
Alpha	↑↓ a)	50	20–50 μV	5 mg morphine in 10 ml normal Saline 0.9%
	↑ b)	20	approx 2 Hz	
	↓ c)	10		
	0 d)	20		
	e) ↓ DF	20		
Beta	0	0	0	
Delta / Theta	↑	33	In individual cases 80 μV	

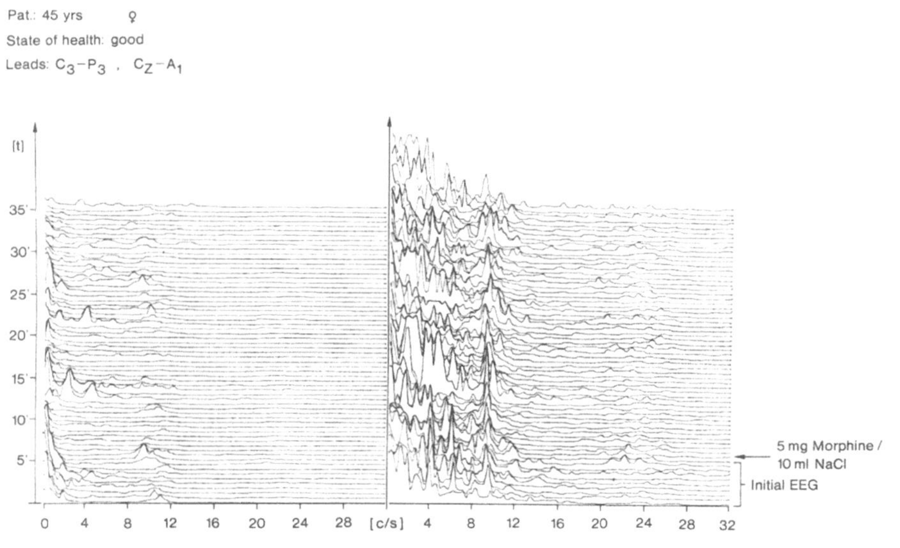

Fig. 7. EEG spectral analysis prior to and after peridural injection of 5 mg morphine in 10 ml normal saline solution. After morphine and a latency period of approximately 5 min, the pre-existing 10 Hz initial alpha EEG is suppressed. Afterward, the alpha activity is alternately reactivated or again suppressed. At the same time, there are considerable increases of activity in the delta/theta range. Clinically no sedation.
Recording conditions: TC, 0.3 s; filter, 70 Hz; calibration, 50 μV = 7 mm; paper speed, 30 mm/s; FFT, 30-s periods

30 min) (Table 5; Fig. 7 [17]). After peridural morphine application, seven of ten patients presented EEG changes in terms of reduced central nervous function. In one case, there were changes that corresponded with an increase in vigilance. Frequently (50% of cases), alterations in electric power in the spectrum of the initial frequency were observed; more rarely (20% of cases), increases in activity were seen in the delta/theta range.

Review of Other Findings. Specific studies on the EEG pattern after peridural morphine application have not been performed. DE CASTRO and LECRON [6] describe occasional EEG signs of hypervigilance as incidental findings after peridural fentanyl injection. Our own findings resemble those under systemic morphine application; they suggest central nervous depression. WIKLER [33] observed signs of hypervigilance (overalertness) in systemic morphine application. However, these signs were not reflected in the EEG. We also considered the data of ENGQUIST [8], who found high morphine concentrations in the liquor space after peridural application, as well as low cerebral levels of morphine after systemic application (0.1% of the given dose (STANLEY 1980). These data suggest that in peridural and systemic application, the drug-induced EEG changes are similar.

Clinical Evaluation. Peridural morphine application leads to the induction of spontaneous sleep in a few patients. More frequently, the level of consciousness is unaffected. Side effects within the 1st h after morphine injection were not observed in the patients studied.

Interpretation of EEG Changes and Clinical Findings. The occurrence of EEG changes does not correlate with clinical effects. A possible explanation could be that the central nervous emergence of the drug was not complete over the test period. With the above-mentioned long circulatory duration [4–6] of the cerebrospinal fluid (CSF), changes in neuronal function could occur that were not manifested clinically.

Summary of EEG Findings with Analgetics for Postoperative Management of Pain (Table 6)

At present, a number of potent analgetics are available for the postoperative management of pain whose analgesic efficiency is known. EEG changes are seen when examining the above-mentioned drugs in equipotent doses. These changes do not reflect the specific analgesic effects, thus ruling out any determination of the analgesic potency of the substance or of the variations within individuals. Analgetics have at the same time an effect on the level of consciousness; this effect is generally evident in the EEG, and individually its mode and degree can be determined. Analgetics lead to intermittent or sustained decreases in the dominant initial frequency. Simultaneously, patients experience relaxation, calming, and even psychic indifference. With only minimal fluctuations in activity, unpleasant sensations, such as fear, prevail. With analgetic application, the dominant initial frequency may fall by 0.5 to 2 Hz in such cases, the psychic relaxing effect of the drug is clinically more pronounced. Analgetics lead to a drug- and dose-dependent increase in activity in the lower frequency range be-

Table 6. Summary of major changes in the EEG and in clinical behaviour under analgetics in equipotent doses ($n=78$)

Substance dose	EEG	Clinical behavior[a]	Side effects	Remarks
Morphine 5 mg i.v. or 10 mg i.v. or 5 mg peridurally	Intermittent de- or increase in dominant basic frequency (5–10 μV); dose-dependent increase in delta (theta) activity (5 or 10 μV)	Objective: sedation 0 – (+) or + Subjective: mild calming, or no change	Central: nausea, vertigo – approx. 40% of cases, more pronounced with low dose. Peripheral: circulatory reactions, approx. 5% of cases	Specific analgetic effect not evident in EEG; sedative component, modifying vigilance, expressed in EEG, correlating with clinically observed changes in consciousness
Pethidine 100 mg i.v.	Reduction of dominant basic frequency, approx. 15–40 μV, increase in activity in delta (theta) range (15–40 μV). Narrowing of basic frequency over 50% in higher frequency ranges (with irregular initial EEG)	Objective: sedation + to + + + Subjective: indifference	Central: restlessness, approx. 16% of cases	
Piritramide 15 mg i.v.	Intermittent de- or increase in dominant basic frequency, 15–20 μV, decrease in frequency by 0.5–2 Hz, increase of activity in delta (theta) range, 10–30 μV	Objective: sedation + to + + Subjective: relaxation, well-being	Central: nausea approx. 10% of cases	
Pentazocine 30 mg i.v.	Reduction of dominant basic frequency, approx. 10–30 μV, increase of activity in delta (thelta) range, approx. 10–15 μV, occasionally 70 μV	Objective: sedation + to (+ + +) Subjective: calming, indifference	Central: respiratory depression, nausea, vertigo, approx. 25%	

[a] Degrees of sedation: + mild, + + moderate, + + + deep

tween 2 and 6 Hz. The extent of the changes indicates the clinical degree of sedation. The analgetics described above have substance- and dose-related central side effects in a number of cases, which cannot be determined from the EEG.

References

A. Textbooks and Summaries

Benzer H, Frey R, Hügin W, Mayrhofer O (1982) Lehrbuch der Anästhesiologie, Reanimation und Intensivtherapie, 5th ed. Springer, Berlin Heidelberg New York

Kuschinsky G, Lüllmann H (eds) (1981) Kurzes Lehrbuch der Pharmakologie. Thieme, Stuttgart

Stanley T (1980) Pharmacocinetics of narcotic compounds. Trends in intravenous anesthesia. In: Aldrete JA, Stanley T (eds) Chicago Year Book Medical Publishers, Chicago

B. Papers

1. Andrews HL (1941) Brain potentials and morphine addiction. Psychosom Med 3:399–409
2. Bailey PW, Smith BE (1980) Continuous epidural infusion of fentanyl for postoperative analysia. Anaesthesia 35:1002–1006
3. Behar M, Magora F, Olshwang D, Davidson JT (1979) Epidural morphine in treatment of pain. Lancet I:527–529
4. Booker PD, Wickes RG, Bryson THL, Beddard J (1980) Obstetic pain relief using epidural morphine. Anaesthesia 35:377–379
5. Bullingham RES, McQuay HJ, Moore RA (1980) Unexpectedly high plasma fentanyl levels after epidural use. Lancet 2:1361–1362
6. DeCastro J, Lecron L (1981) Peridurale Opiat-Analgesie. Verschiedene Komplikationen und Nebenwirkungen. In: Zenz M (ed) Peridurale Opiat-Analgesie. Fischer, Stuttgart, p 103–120
7. DeCastro J, D'Inverno E, Lecron L, Levy D, Toppet-Balatoni E (1980) Perspectives d'utilisation de morphinoides en anesthésie loco-regionale justification – premier résultats. Anesth Anal Rean 37:17–24
8. Engquist A (1981) Grundlagen der periduralen Opiat-Analgesie und klinische Erfahrungen. In: Zenz M (ed) Peridurale Opiat-Analgesie. Fischer, Stuttgart, p 1–12
9. Husemeyer RPM, O'Connor C, Davenport HT (1980) Failure of epidural morphine. Anaesthesia 35:161–163
10. Itil T, Fink M (1966) Anticholinergic drug-induced delirium: Experimental modification, quantitative EEG and behavioral correlations. J Nerv Ment Dis 143:492–507
11. Janssen. Scientific information pamphlet on Dipidolar
12. Janssen PAJ (1961) Piritramide (R3365), a potent analgesic with unusual chemical structure. J Pharmacol 13:513–530
13. Janzen R, Keidel WD, Herz A, Steichele A (eds) (1972) Schmerz: Klinische Erfahrungen in der Anästhesiologie mit Pentazocin. Thieme, Stuttgart, p 368–397
14. Kay B (1971) A clinical investigation of piritramide in the treatment of postoperative pain. Br J Anaesth 43:1167–1171
15. Kubicki S (1975) Die Physiologie der zentralen Schmerzverarbeitung. In: Rügheimer E, Heitmann D (eds) Die Neuroleptanalgesie – Bilanz einer Methode. Thieme, Stuttgart, p 3–10
16. Kubicki S, Stölzel R, Haas J (1972) Auslöschung des Fentanyl-Effektes durch Pentazocin im EEG und somatisch-sensorisch evozierte Potentiale. In: Henschel WF (ed) Postoperative Schmerzbekämpfung. Schattauer, Stuttgart, pp 103–110
17. Lips U, Pichlmayr I (1981) Vergleich der zentralen Wirksamkeit periduraler und systemischer Morphin-Gaben an Hand von EEG-Spektralanalysen. Intensivmed Prax 4:51–56

18. Mule SJ (1971) Physiological dispositons of narcotic agonists and antagonists, narcotic drugs: Biochemical pharmacology. In: Clouet DH (ed) Tissue responses to addictive drugs. Plenum, New York
19. Nilsson (1972) Diskussionsbemerkung auf dem 3. Europäischen Anästhesiekongreß 1970 in Prag. In: Hoder J, Jedlicka R, Pokorny J (eds) Advances in anesthesiology and resuscitation, vol 2. Aricenum-Czechoslovak Medical Press, Prague, pp 1573–1576
20. Pichlmayr I, Lips U (1979) Pethidin-Effekte im Elektroenzephalogramm. Anaesthesist 28:433–442
21. Piepenbrock S, Zenz M, Otten G (1981) Peridurale Opiat-Analgesie in der postoperativen Phase. In: Zenz M (ed) Peridurale Opiat-Analgesie. Fischer, Stuttgart, pp 47–62
22. Rifat K (1972) Pentazocin in der sequentiellen analgetischen Anästhesie. Br J Anaesth 44:175–182
23. Saarne A (1969) Clinical evaluation of the new analgesic piritramide. Acta Anaesthesiol Scand 13:11–19
24. Scheider J, Remond DA (1949) Notes préliminaires concernant l'action de la morphine à doses variables sur le trancé EEG. Rev Neurol (Paris) 81:512–513
25. Scott DB, McClure J (1979) Selective epidural analgesia. Lancet 1:1410–1411
26. Scott PV, Bowen FE, Cartwright P, Mohanrav BC, Deely D, Wohterspoon HG, Sumreim IMA (1980) Intrathecal morphine as sole analgesic during labour. Br Med J 281:351–355
27. Sertürner FH (1817) Über das Morphium eine neue salzfähige Grundlage und die Mekonsäure als Hauptbestandteil des Opiums. Gilbert's Ann Physik 55:56–89
28. Synder SH (1975) Opiate receptor in normal and drug altered brain function. Nature 257:185–189
29. Sonntag H, Meyer-Burgdorff C (1972) Postoperative Schmerzbehandlung mit Piritramid. In: Hoder J, Jedlicka R, Pokorny J (eds) Advances in anaesthesiology and resuscitation, vol 2. Avicenum-Czechoslovak Medical Press, Prague, p 1480–1481
30. Takki S, Tammisto T (1973) A comparison of pethidine, piritramide, and oxycodone in patients with pain following cholecystectomy. Anaesthesist 22:162–166
31. v. d. Berg B, v. d. Berg E, Zenz M (1981) Peridurale Morphin-Analgesie und Sympathikusblockade. In: Zenz M (ed) Peridurale Opiat-Analgesie. Fischer, Stuttgart, pp 31–46
32. Wang JK, Nauss EA, Jürgen ET (1979) Pain relief by intrathecally applied morphine in man. Anesthesiology 50:149–151
33. Wikler A (1954) Clinical and electroencephalographic studies of the effects of mescaline. N-allgenormorphine and morphine in man. J Nerv Ment Dis 120:157–175
34. Winthrop. Scientific information pamphlet on Fortral
35. Zenz M (1981) Peridurale Opiat-Analgesie. Dtsch Med Wochenschr 16:482–485
36. Zenz M, Piepenbrock S, Hüsch M, Otten G, Otten B (1981) Peridurale Opiat-Analgesie. Anaesthesist 30:28–30
37. Zenz M, Piepenbrock S, Hüsch M, Schappler-Scheele B, Neuhaus R (1981) Erfahrungen mit längerliegenden Periduralkathetern. Peridurale Morphin-Analgesie bei Karzinompatienten. Regional Anaesthesie 4:26–28
38. Zenz M, Piepenbrock S, Otten B, Otten G, Neuhaus R (1981) Peridurale Morphin-Analgesie I. postoperative Phase. Anaesthesist 30:77–80
39. Zenz M, Piepenbrock S, Otten B, Otten G (1980) Epidurale Morphin-Injektion zur Schmerzbekämpfung. Fortschr Med 98/99:306–308

VII. Anesthesiological and Operative Procedures with Potential Cerebral Effects

CONTENTS

In addition to anesthesia, further anesthesiological procedures accompanying surgery, the effects of surgery, and operations involving particular stress on the circulation, or phases of such operations, have consequences on the brain, which in the EEG can be determined by large alterations in cerebral function. As always, the reactions of the EEG are fairly stereotype, i.e., various injurious agents lead to similar EEG changes; knowledge of the clinical circumstances is, therefore, required for an evaluation of their significance. The following basic rule can thus be applied: Changes in the EEG in terms of a decelerating frequency and a leveling off of the voltage during the steady state of anesthesia under unchanging clinical conditions should always be taken as a warning signal of a deficient supply of oxygen to the brain. The cause must be detected and treated [13]. Changes toward a deceleration of frequency suggest a reduction in the level of anesthesia, either due to metabolism of the anesthetic or increased surgical pain stimuli – as long as other clinical parameters cannot explain such changes.

Anesthesiological Procedures

Manipulations After Induction of Anesthesia. Endotracheal intubation, endotracheal suction, insertion of a gastric tube, and the setting up of a central venous catheter (cvp line) are obligatory measures prior to medium or extensive surgery. These measures lead to an abrupt reduction in the level of anesthesia, especially if shortly after induction of anesthesia a steady state in stage III_2 has not been achieved. Marked phenomena of desynchronisa-

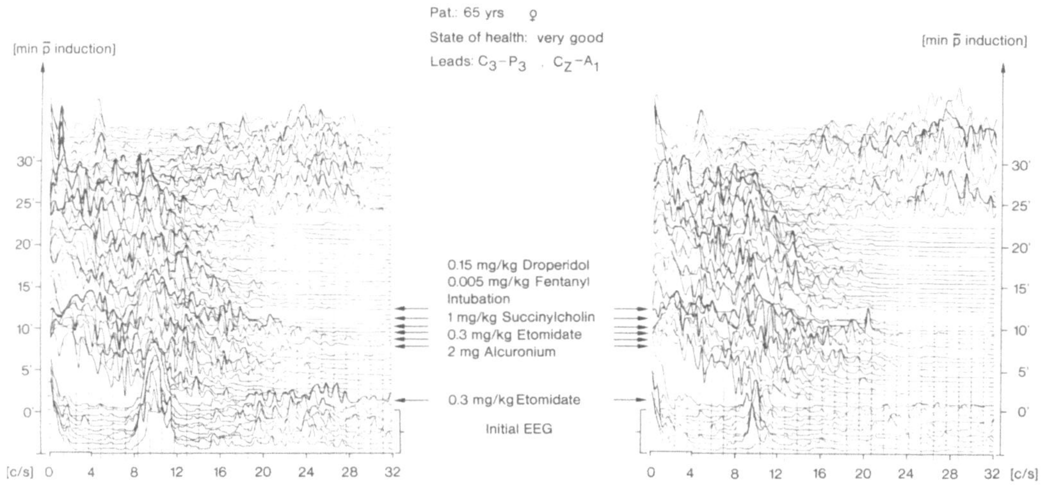

Fig. 1. Arousal reaction as a result of the insertion of a central-venous catheter 20 min postinduction with etomidate-induced neuroleptanalgesia.
Recording conditions: TC, 0.3 s; filter, 70 Hz; calibration, 50 μV = 7 mm, paper speed, 30 mm/s; FFT, 30-s periods

tion are seen in the EEG, occasionally accompanied by clinical motor reactions (Fig. 1). In patients with preexisting severe cerebral damage, intubation or endotracheal suction can occasionally lead to signs of increased anesthesia, with an increase in slow high waves. Hypoxic influences on the ganglion cells are regarded as the cause (Sadove et al. 1967).

Intravenous Fluid Schemes. Infusions of electrolyte or colloidal solutions, protein fractions, or whole blood did not lead to any alterations in the brain function in our own studies. Other authors report that large amounts of cold infusion or transfusion solutions may cause a decrease in frequency in the EEG, probably due to cerebral hypoxia (Sadove et al. 1967).

Drugs. The intraoperative application of antibiotics, cardiomimetic drugs, and catecholamines does not lead to visible changes in the EEG activity. In individual cases, even in deep stages of anesthesia, marked signs of arousal occur after intravenous drug application, here seen under continuous peridural ethrane anesthesia, following application of calcium ions (Fig. 2). This surprising reaction could not be explained.

Effects of Surgery

Pain Stimuli. Empirically, during the course of abdominal surgery, mechanical traction on the peritoneum and stomach usually cause a clinical

Pat.: 78 yrs ♀

State of health: satisfactory

Leads: C_3–P_3 , C_Z–A_1

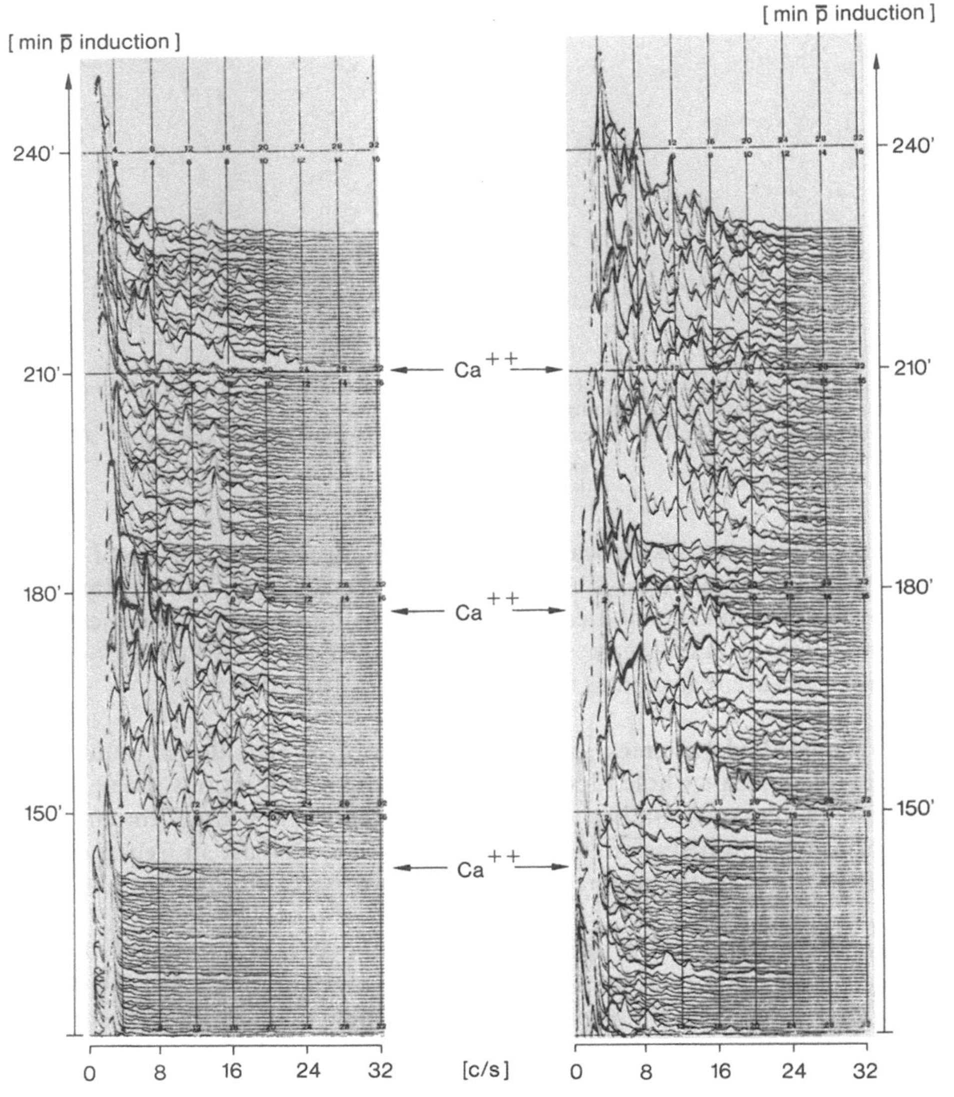

Fig. 2. EEG spectral analysis in the steady state of deep enflurane anesthesia. Injection of ionized calcium leads to sudden activation (including the faster bands) in the EEG. Recording conditions: TC, 0.3 s; filter, 70 Hz; calibration, 50 µV = 7 mm; paper speed, 30 mm/s; FFT, 30-s periods

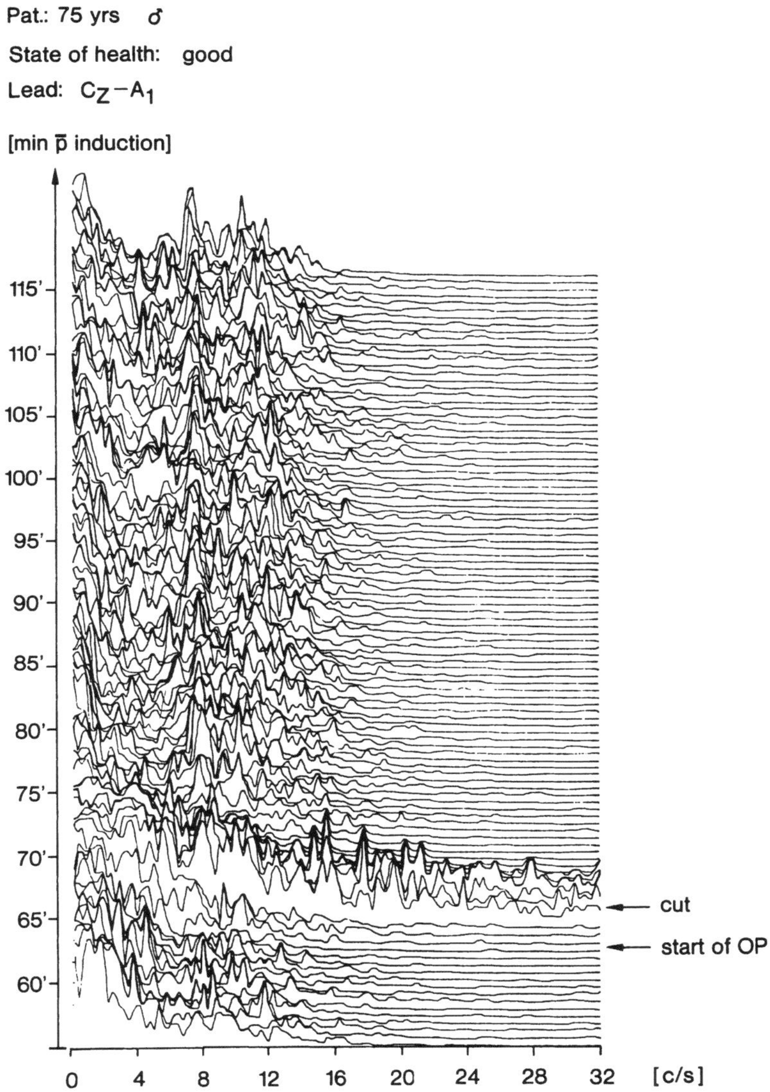

Fig. 3. EEG pattern of a barbiturate-induced neuroleptanalgesia – analgetic phase. Pain stimulus by the surgical cut leads to an arousal reaction with activities over the entire frequency band.
Recording conditions: TC, 0.3 s; filter, 70 Hz; calibration, 50 μV = 7 mm; paper speed, 30 mm/s; FFT, 30-s periods

phase of awakening with singultus, which is occasionally accompanied by reactions of the cardiovascular system. In the EEG, the decrease in the level of anesthesia is evident as an increase in frequency (Fig. 3) (SADOVE et al. 1967). In deep or very deep stages of anesthesia, clinical and encephalographic reactions fail to occur.

Intraoperative Hypotensive Crisis. Profuse intraoperative hemorrhages rapidly lead, via a general circulatory reaction, to a cerebrally deficient circulation, with a reduction in frequency and voltage in the EEG (Fig. 4). The critical limit, with a systolic blood pressure of approximately 70 mm Hg in patients with a healthy vascular system, is markedly higher than with a slow

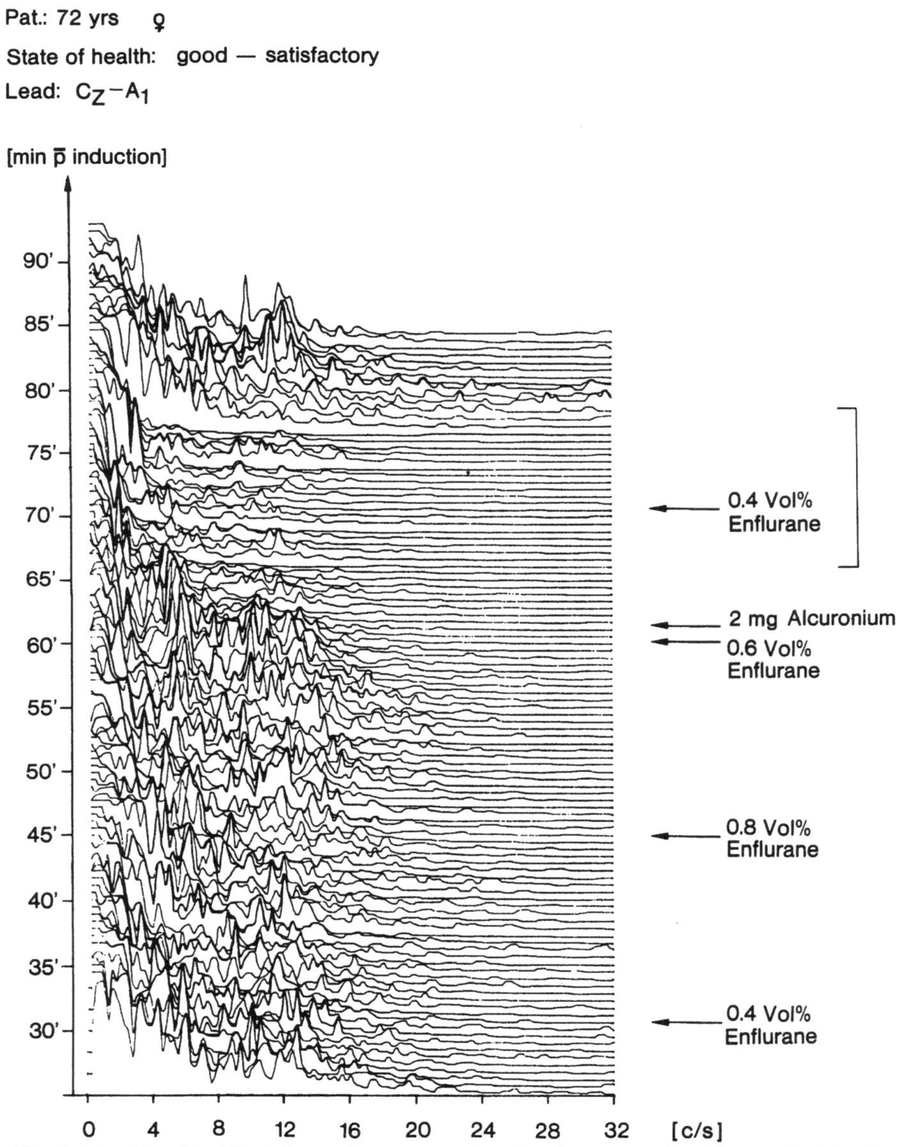

Fig. 4. Considerable EEG depression as a result of an intraoperative decrease in blood pressure for 10 min down to 70/30 mm Hg (anesthesia – combined PDA-enflurane anesthesia). Recording conditions: TC, 0.3 s; filter, 70 Hz; calibration, 50 μV = 7 mm; paper speed, 30 mm/s; FFT 30-s periods.

decrease in blood pressure [1, 2, 6]. Vascular surgical manipulations on the carotids involving vasal compression may provoke a local decrease in frequency in individual cerebral areas. In the case of a locally insufficient blood supply with the corresponding frequency changes, increases in frequency are observed postoperatively after the circulation has improved (Brechner et al. 1962; Sadove et al. 1967).

Extracorporal Circulation (ECC). In thoracic and cardiovascular surgery, especially in operations involving extracorporal circulation, continuous EEG monitoring has, for many years, proven to be a valuable aid in evaluating cerebral perfusion and, thus, at the same time, the performance of the pump-oxygenator [8–11] (Brechner et al. 1962; Kugler 1966, Sadove et al. 1967). EEG recording as a monitor of cerebral integrity requires the steady state of light anesthesia with an almost constant basic frequency, fast low waves as background, and occasional theta spindles as a manifestation of deep phases of sleep. When lowering the internal body temperature, as is commonly performed in open heart surgery with circulatory standstill, the temperature does not fall below 28°–30° and thus has little influence on electric functional signs.

In the course of thoracic or cardiac surgery, cerebral perfusion during the various stages of surgery may be impaired by different causes. Even during thoracotomy, an inadequate mechanical ventilation with transient hypoxia is possible due to retraction of the lungs (e.g., in patients with a limited respiratory reserve). Frequency decelerations in the EEG would then be an early warning signal. Exposure of the heart and large vessels may lead to partial obstruction of the superior vena cava due to mechanical pressure of the retractors or cardiac torsion.

This may result in a decreased total cardiac output. Subsequent venous congestion in the brain or decrease in the total cerebral perfusion also lead to signs of cerebral hypoxia, with a decrease in frequency cannulation of the large vena cava – necessary in ECC, regularly provokes a fast reversible decrease in frequency in the EEG without giving rise to simultaneous circulatory or cardiac reactions. Transient increases in intracranial pressure or reflex mechanims are thought to be causative factors [7, 8, 11].

The beginning of extracorporal perfusion involves a brief cerebral cessation of perfusion with a decrease in frequency in the EEG. Again, several factors, such as circulatory depression, transient increased cerebral venous pressure, or influences due to the influx of cold perfusion solution, have been considered as causes. An unaltered EEG can be expected with technically perfect extracorporal perfusion. The critical blood pressure limit for adequate cerebral perfusion in this phase is in the range of approximately 30–40 mm Hg, as opposed to 70–75 mm Hg under normal conditions. The lower pressure level is explained by the constant volume output of the heart-lung machine. While decreased pressure due to cardiac causes is also accompanied by a decrease in volume output, an adequate perfusion is guaranteed with ECC despite a lower pressure. Perfusion errors and individual differences in the tolerance toward a low blood pressure manifest themselves electroencephalographically as a decrease in frequency and increase in amplitude (or a decrease in the latter), depending on the degree of the impairment. Signs of inadequate central nervous perfusion necessitate the immediate detection and elimination of the hazardous cause. Prior to the final transition from extracorporal perfusion to proper cardiac drive, a phase of partial bypass must be overcome. The resulting coexistence of machine- and cardiac-driven perfusion gives rise to mild changes in the EEG. In the postbypass period EEG changes of various degrees are possible due to alterations in cardiac action together with circulatory fluctuations; these changes indicate the actual state of the cerebral supply. EEG changes can be reversed after their cause has been eliminated, depending on how long the factor was active. The time needed for complete cerebral regeneration according to the EEG may indicate cerebral damage in the postoperative period that could manifest itself clinically [4, 5, 8, 11, 12] (BRECHNER et al. 1962).

Evaluation of EEG Effects Due to Anesthesiological Procedures and Surgery

The EEG as an additional monitoring system during cerebrally hazardous surgery or hazardous anesthesia indicates changes in cerebral function. Evaluation of the EEG leads to early disclosure of any risks and their elimination can increase the safety of the patient. However, for technical reasons, the routine use of EEG monitoring during uncomplicated anesthesia or surgery is debatable [3].

References

A. Textbooks and Summaries

Brechner VL, Walter RD, Dillon JB (1962) Practical electroencephalography for the anesthesiologist. Thomas, Springfield

Kugler J (1966) Elektroenzephalographie in Klinik und Praxis, 2nd edn. Thieme, Stuttgart

Sadove MS, Beck AD, Gibbs FA (1967) Electroencephalography for anesthesiologists and surgeons. Pitman, London

B. Papers

1. Bellville JW, Artusio JF Jr (1956) Effect of Arfonad® on anesthetic requirements during Cyclopropane anesthesia. Anesthesiology 17:347–352
2. van Bergen FH, Bucktey JJ, French LA, Dobkin AB, Brown JA (1954) Physiologic alterations associated with hexamethonium-induced hypotension. Anesthesiology 15:507–536

3. Coons RE, Keats AS, Cooley DA (1959) Significance of electroencephalographic changes occurring during cardiopulmonary by-pass. Anesthesiology 20:804–810
4. Davenport HT, Arfel G, Sanches FR (1959) The electroencephalogram in patients undergoing open heart surgery with heart-lung by-pass. Anesthesiology 20:674–684
5. Kavan EM, Brechner VL, Walter RD, Linde LM (1959) Electroencephalographic and electrocardiographic patterns during open heart operations with the use of cardiopulmonary bypass. Can Anaesth Soc J 6:356–364
6. Kubicki S, Just O (1957) Das hirnelektrische Bild bei extremer künstlicher Blutdrucksenkung. Anaesthesist 6:143–147
7. Kubicki S, Just O (1959) Das EEG im Verlauf von Herzoperationen mit Kreislaufunterbrechung. Anaesthesist 8:1–5
8. Kubicki S, Trede M, Just O (1960) Die Bedeutung des EEG bei Herzoperationen in Hypothermie und bei extrakorporaler Zirkulation. Anaesthesist 9:119–123
9. Martin JT, Faulconer H, Bickford RG (1959) Electroencephalography in anesthesiology. Anesthesiology 20:359–376
10. Owen SG, Adams JE, Dawson BE, Lance EM, Sawers JL, Scott HW (1958) Observed central nervous system responses during experimental employment of various pump oxygenators. Surgery 44:240–254
11. Trede M, Kubicki S, Just O (1959) Über EEG-Beobachtungen bei Herzoperationen mit extrakorporalem Kreislauf. Anaesthesist 8:76–82
12. Walter RD, Kavan EM, Brechner VL, Maloney JV (1958) EEG-changes during cardiac surgery with cardiopulmonary by-pass. Electroencephalogr Clin Neurophysiol 10:180
13. Wiemers K, Puppel H (1960) Praktische Bedeutung der EEG-Registrierung im Operationssaal, Kreislaufmessungen. Abstracts of the 2nd Freiburg Colloquium on circulatory measurement, vol 2, p 21–34

VIII. Perioperative Disturbances with Cerebral Effects

CONTENTS

Disorders of the physiological balance result in EEG changes. The EEG has proven to be a highly sensitive indicator of alterations in cerebral function. Since various noxae may cause similar encephalographic phenomena, the EEG allows an evaluation of the extent and degree of cerebral alteration. Conclusions as to the actual cause are not possible from the EEG alone but can be made in conjunction with the clinical information. The course of EEG changes due to clear-cut alterations in cerebral perfusion and metabolism has been studied and is known to a certain extent.

Anoxia – Hypoxia (Table 1)

[10, 11, 13, 16, 20, 22, 23] (BRECHNER et al. 1962; CHRISTIAN 1975; GIBBS and GIBBS 1958, JUNG 1953; KUGLER 1981; NEUNDÖRFER and WITZEL 1975)

Cerebral oxygen depletion leads to a rapid progressive depression of electroencephalographically determinable function after very brief activation. After short periods of anoxia, the resulting functional standstill terminates in irreparable morphological damage. The oxygen saturation level of the blood is, with individually varying values, between 60% and 95%. Under hypoxia, depending on the initial value, EEG changes can be seen with oxygen saturation levels of 40%–70%; these changes occur prior to, or at the same time as, clinically detectable cyanosis and – in alert patients – before loss of consciousness. With simultaneous hypocapnia, the EEG signs due to hypoxia are initially masked; they thus become noticeable somewhat later than with normocapnia. This may be due to the well-known increase in cerebral perfusion with hypocapnia. With slowly increasing hypoxia, the amplitude of the alpha activity is initially raised. This is followed by an increase in frequency (beta activity of high amplitude), a reduction in frequency (delta/theta activity of high amplitude), a reduction in amplitude (low delta activity), and finally an isoelectric EEG. With cardiac arrest and cerebral anoxia, the symptoms of primary desynchronization are frequently passed through so rapidly that only the occurrence of synchronization or even of the isoelectric phases is noticed.

Acute anoxia within 20 to 30 s leads to a failure in the function of the ganglion cells. Local hypo- or anoxia cause simultaneous, yet locally circumscribed, EEG symptoms, which are not generally evident from the scalp lead signals. If a sufficient oxygen supply is resumed within 3 min, the EEG returns to normal with a steady increase in voltage and frequency after a latency of a few minutes. With further sufficient oxygen supply and stable circulatory conditions, no neurological deficits should be expected. The fact that a 3-min cerebral anoxia can be survived without sequelae but that this is also the time-limit of tolerance for cerebral ischemia was observed and reported as early as 1857 by Brown-Sequard [6] and has since been confirmed both in animal experiments [27, 38, 39] and clinically [1, 11]. Under hypothermic metabolic conditions, this limit of tolerance can be increased to 7½ min. The time taken for the EEG to recover a normal pattern provides indications for the prognosis. Regeneration of the electric potentials within 30 min is held to be prognostically promising. Longer recovery times indicate residual damage. Sustained isoelectric phases with no tendency to shift into lower grades of cerebral depression and the "file pattern" (low very fast frequencies of 30 Hz) are regarded as agonal or poor prognostical signs.

Generally, the chance of cerebral recovery after anoxic or ischemic damage also depends on the preoperative parameters. The compensatory mechanisms that are available are so slight after such events that further even minor deviations in the clinical parameters immediately deteriorate cerebral function and, thus, the EEG. The following is the clinical pattern with regard to recovery of the brain after a severe impairment of oxygen supply: The return of continuous EEG activity within 30 min and a short recovery time of the EEG signs up to mild cerebral depression suggest a good prognosis. The onset of continuous electric activity after a period of 30 min, the absence of recovery signs in the EEG, and the occurrence of the "file pat-

Table 1. EEG changes with perioperative hazards and metabolic disorders

Pathophysiological disorder	EEG changes	Remarks
Hypotension		
Mean arterial pressure 75 mm Hg	None – due to compensatory mechanisms (autoregulation)	EEG reveals alterations only when circulatory cerebral decompensation is impending (late warning system). With a history of cerebral sclerosis, the critical mean arterial pressure is over 120 mm Hg
Mean arterial pressure 50–70 mm Hg	Frequency deceleration, increase in voltage	
Mean arterial pressure 40 mm Hg	Further decrease in frequency, decrease in voltage	
Hypoxia		
Mild	Frequency acceleration, increase in voltage	EEG useful as "early warning" system
Moderate	Frequency decrease with increase in amplitude	
Severe	Further decrease in frequency with decrease in amplitude	
Anoxia	Cessation of cortical activity (iso-electric EEG)	
Hypothermia		
28 °C	None	
28 °–24 °C	Increase in frequency, followed by frequency decrease, decrease in voltage,	With severe hypothermia, the EEG loses significance in monitoring further hazards
24 °C	Marked decrease in frequency	
15 °C	Disappearance of measurable cortical activity (isoelectric EEG)	
Hypercapnia		
Mild	Frequency acceleration	Easily confused with signs of increasing anesthesia in the EEG
Moderate	Decrease in frequency	
Severe	Loss of activity	
Hypocapnia	Increase in cortical excitability, slow high waves	In the case of previous seizures disorders are triggered
Hyperglycemia	None	
Hypoglycemia		
Mild to moderately severe	Large slow waves, reduction in voltage, loss of activity	Simultaneous loss of consciousness, possible provocation of paroxysmal EEG activities, possible clinical cerebral seizures
Hyperthyroidism	Alpha frequency increase approx. 2 Hz, beta components	
Hypothyroidism	Alpha frequency decrease approx. 2 Hz, delta/theta components	
Hypocalcemia	None, occasional beta activity, occasional paroxysmal activity	
Cushing's disease	Increase in beta activity	No proven association

Table 1. (continued)

Pathophysiological disorder	EEG changes	Remarks
Menstrual cycle		
Ovulation	Increase in alpha frequency approx. 2 Hz Increase in beta activity	EEG effect may not be present
Menstruation	Decrease in alpha frequency approx. 2 Hz Increase in delta activity	
Coma through metabolic decompensation		
Hepatic coma	Slow irregular waves	
Uremic coma	Reduction of voltage	
Chronic hypoxidosis		
Mild	Decrease in alpha frequency approx 2 Hz Increase in beta activity	
Moderate	Additional delta/theta activity	
Severe	Delta/theta frequency with high voltage	

tern" after reestablishment of circulation and sufficient oxygenation, are prognostically unfavorable. Repeated cardiac arrests with intermediate periods of sufficient cerebral oxygenation do not aggravate the prognosis for the reestablishment of cerebral functions as a result of intervening recovery time.

With increase in the depth of anesthesia as far as overdose, the EEG changes are similar to those seen with hypoxia. Hazards for cerebral oxygen supply cannot be recognized from the EEG at this stage of anesthesia.

Hypotension (Table 1)

[14, 23, 36] (Brechner et al. 1962; Christian 1975; Gibbs and Gibbs 1958; Jung 1953; Kugler 1981)

Since 1938 it has been known that a decrease in blood pressure causes EEG changes [2]. The cerebral perfusion remains constant over a wide range during blood pressure changes due to "cerebral autoregulation." A well-functioning vasculature is required for an adequate response of the autoregulation. The autoregulation is impaired or may fail when the cerebral vessels lose their reactivity due to sclerosis or maximal dilation or narrowing. With healthy arteries, at systolic pressures of between 50–70 and 220 mm Hg, the cerebral perfusion and thus the oxygen supply to the brain remain intact. With further decreases in blood pressure, the cerebral perfusion declines proportionally to the systemic pressure and eventually becomes inadequate. The EEG findings coincide with these changes. Thus, with decreased blood pressure, the EEG detects warning signs only when there is decompensation in the circulatory supply to the ganglion cells. Since in the case of previous cerebral damage or cerebral arteriosclerosis in the older age-group the compensatory reserve for cerebral vasoregulation is decreased and with considerably higher blood pressure values (compensatory hypertension) symptoms of decompensation have to be expected, the EEG here can indicate the point when a decrease in blood pressure becomes dangerous for the brain.

When lowering the blood pressure by ganglionic blockage, arterial mean pressures of 40 mm Hg are tolerated for 60 min without EEG changes; a rapid fall in blood pressure down to the same reading exceeds – even under ganglionic blockage – the critical limit of vascular compensation. An inadequate blood supply to the brain, as a result of lowered blood pressure, leads to changes in the electric activity similar to those seen with hypoxia. Here, without any additional narcotic medication, EEG signs indicative of an increase in the depth of anesthesia are seen. According to the extent of the deficiency and the resultant cerebral depression, slow waves are recorded in the unanesthetized patient, which are reduced in voltage and are transformed into burst-suppression phases and – under the extreme conditions of a cardiac arrest – into an isoelectric EEG. Thus, the first EEG changes are to be expected in patients with a healthy vasculature at systolic blood pressures of 50–60 mm Hg, and a loss of voltage at readings of 45–50 mm Hg. In cases of local ischemia, the corresponding EEG signs are locally confined; depending on the localization, they cannot be reliably recorded. In cardiac arrest, the efficiency of cardiac massage may also be determined by the recovery signs in the EEG. Thus, the EEG does not show early warning symptoms of all decreases in blood pressure, but it may indicate when the limit of safety for cerebral circulation has been exceeded.

Hypothermia (Table 1)

[4, 19, 22] (Brechner et al. 1962; Christian 1975; Gibbs and Gibbs 1958; Jung 1953; Kugler 1981; Neundörfer and Witzel 1975)

Hypothermia as a result of accidents in the mountains or in water and hypothermia in alcohol-intoxicated subjects brought about by sleeping outside at night has a protective effect against concomitant parenchymal damage by mildly decreasing the temperature of the body care. Therapeutically, low temperatures are used to avoid tissue damage in the presence of ischemia. Such techniques are mainly applied in cardiac surgery, neurosurgery, and intensive care. At present, the therapeutic temperature is usually adjusted to around 28 °C; further cooling is possible, especially in cardiac surgery in children.

Without pharmacological assistance, cerebral functions are not affected down to a body temperature of 28–30 °C. The necessary additional neuroleptic medication reveals slow high-voltage EEG activities at the beginning of a deliberately induced hibernation as well as – in the steady state – fast low-voltage activities in addition to the baseline frequency as a manifestation of the cerebral drug effect; these changes are not due to effects of the temperature decrease. With temperatures below 28 °C, the progressive hypothermia with corresponding metabolic impairment results in cerebral depression, which is evident in the EEG. Between 24 ° and 28 °C, there is initially a decrease in voltage and, after a brief increase in frequency, a frequency decrease. Lower temperatures lead to a very marked deceleration of frequency and voltage reduction down to isoelectric patterns at about 15 °C. In rewarming, the EEG stages corresponding to the stages of hypothermia are passed through reverse order. Return of consciousness coincides with beta-activity of normal amplitude in the EEG.

Acid-Base Imbalance (Table 1)

[10, 33, 35] (Brechner et al. 1962; Christian 1975; Gibbs and Gibbs 1958; Jung 1953; Kugler 1981; Neundörfer and Witzel 1975)

Moderate and severe respiratory and metabolic disorders of the acid-base balance influence cerebral parameters. The changes in CO_2 blood levels have definite and exactly determinable effects on cerebral perfusion: Increases in carbon dioxide lead to cerebral vascular dilatation with increased perfusion, decreases in carbon dioxide lead to vascular constriction with decreased perfusion. The changes in perfusion are linear and can be calculated in CO_2 levels of between 20 and 80 mm Hg. Concomitant functional changes in the brain are probably not exclusively based on the circulatory effects of acidosis or alkalosis. Hypocapnia with alkalosis increases cortical excitability; in the neurological diagnosis of epilepsy, this is used as a stress test when searching for pathological EEG changes, which then become more distinct. With a considerable decrease in CO_2, high slow frequencies become apparent in the EEG, disappearing under normocapnia within a few seconds or minutes. Long-lasting alkalotic respiratory or metabolic imbalance causes diffuse abnormalities of the encephalographic pattern, with spike-wave (sw) complexes and changes similar to those seen in epileptics. Increased CO_2 first leads to an accelerated frequency, then to a decrease in frequency and loss of voltage. In anesthesia both by hyper- and hypocapnia, signs of increased depth of anesthesia become visible in the EEG, i.e., a shift toward slow regular rhythms. In anesthesia, the EEG represents a late warning sign of hypocapnia, e.g., due to complications in artificial respiration. Further clinical signs, such as increased tidal volume with spontaneous breathing, increase in blood pressure and pulse rate, and cardiac arrhythmias precede the EEG changes. The effects of transitory disorders of the acid-base balance on the EEG are reversible.

Metabolic, Electrolytic, and Hormonal Imbalance (Table 1)

[3, 5, 7–10, 12, 15, 17, 18, 21 24–26, 28–32, 34, 37, 40–42] (BRECHNER et al. 1962; CHRISTIAN 1975; DUMERMUTH 1972; GIBBS and GIBBS 1958; JUNG 1953;, KUGLER 1981; NEUNDÖRFER and WITZEL 1975)

Hypo- and Hyperglycemia. Hyperglycemia (over 200 mg% = 11 mmol/l) usually does not affect the EEG, and even has positive effects on seizures disorders (decrease in signs of petit mal). On the other hand, hypoglycemic metabolic changes – according to severity – lead to mild or severe cerebral depression. This is usually reversible; it may, however, cause residual symptoms when sustained longer. In the EEG, the alpha activity is initially reduced. With blood sugar readings of below 60 mg% (approx. = 3.3 mmol/l), high-voltage slow waves appear almost simultaneously with the loss of consciousness; with a further decrease in blood glucose, they pass through a stage of reduced voltage and disappear. With severe hypoglycemia, cortical activity can no longer be determined. Occasionally, during hypoglycemic fluctuations in blood sugar levels, cerebral seizure patterns are also provoked, which can remain as residual sequelae of hypoglycemic metabolic changes.

Hypo- and Hyperthyroidism. Changes in thyroid function have a particularly marked effect on the background activity. In alpha frequency under normal metabolic conditions of 8–10 Hz is reduced by hypothyroidism to 7–10 Hz, according to the degree of generally reduced metabolism. Delta/theta components and reductions in voltage frequently occur as concomitant symptoms. In hyperthyroidism, alpha activity is accelerated: Basic frequencies of 12–13 Hz result; in addition, low beta waves (12–13 Hz) are activated.

Hypocalcemia (Tetany). The resting EEG frequently remains unchanged in hypocalcemic tetany; occasionally, beta waves are activated. Series of delta waves – frequently activated by hyperventilation – as well as arcade patterns are seen in the EEG as a result of a tetanic seizure. Parathyroprival tetany, which may occur as a consequence of total strumectomy, causes paroxysmal graphoelements, which disappear with substitution therapy.

Impaired Adrenal Function. In patients with Cushing's disease, increased beta activity is observed (52–80% of cases) compared with normal subjects; no conclusive correlation between the endocrine situation and the beta activity has been established.

Menstrual Cycle. The hormonal rhythms during the female menstrual cycle lead to frequency increases at the time of ovulation (12–13 Hz and beta activation) and to frequency deceleration prior to the onset of menstruation (7–10 Hz, delta waves). The EEG may also be unaffected.

Hepatic Disease. With viral hepatitis, cerebral infection sites can be detected from the corresponding neurological symptoms along with the simultaneous local EEG changes. Chronic liver disease generally leads to a leveling off of the previous frequencies – this is regarded as unfavorable with respect to cerebral integrity. In hepatic coma, slow irregular activities typify the pattern.

Renal Disease. The EEG is unaffected by the compensatory mechanisms counteracting the anemia present in chronic renal disease. Uremic coma leads to marked frequency decelerations; the wave pattern is irregular.

Chronic Hypoxidosis. Chronic hypoxidosis as a result of cardiovascular or pulmonary disease gives rise to an irregular EEG with both slow and fast activities. Initially, there is a deceleration of the basic frequency. Subsequently, there is an increase in beta activity, followed later by additional theta or delta waves. In the case of a chronic cerebrally deficient supply, an irregular EEG results (delta to beta), with no dominant frequency range. The formation of a delta/theta pattern indicates the stage of an advanced chronic cerebrally deficient supply.

Summary of EEG Findings with Perioperative Disturbances

EEG monitoring is of diagnostic and therapeutic value in anesthesiology in cases where there is the danger of decompensation of the cerebral perfusion and metabolic situation. Cerebral signs of depression always occur when autoregulation or other compensatory mechanisms for the maintenance of the physiological cerebral balance reach their limits. Since in the case of any significant history of illness – especially of the cerebral circulation – very wide individual differences occur within the limits of compensation, the EEG allows conclusions to be made with regard to the onset of a cerebral hazard for the individual patient. Beyond this, the degree of cerebral impairment through various factors and – after removal of the noxious agent – the course of the recovery can be determined from the encephalographic parameters. After severe damage, the pattern and degree of recovery signs in the EEG are prognostic indicators. Similarly, a preoperatively recorded EEG can provide useful information for the anesthesiologist with regard to any existing impaired cerebral functions.

References

A. Textbooks and Summaries

Brechner VL, Walter RD, Dillon JB (1962) Practical electroencephalography for the anesthesiologist. Thomas, Springfield

Christian W (1975) Klinische Elektroenzephalographie, 2nd edn. Thieme, Stuttgart

Dumermuth G (1972) Elektroenzephalographie im Kindesalter. Thieme, Stuttgart

Gibbs FA, Gibbs EL (1958) Atlas of electroencephalography vols I–III, 2nd edn. Addison-Wesley, Reading MA

Jung R (1953) Das Elektroenzephalogramm (EEG). In: Handbuch für Innere Medizin, vol IV/1. Springer, Berlin Heidelberg New York

Kugler J (1981) Elektroenzephalographie in Klinik und Praxis, 3rd edn. Thieme, Stuttgart

Neundörfer B, Witzel K (eds) (1975) EEG-Fibel. Fischer, Stuttgart

B. Papers

1. Bellville JW, Artusio JF, Glenn F (1955) The electroencephalogram during cardiac manipulation. Surgery 38:259–271
2. Berger H (1938) Über das Elektroenkephalogramm des Menschen XIV. Mitteilung. Arch Psychiatr Nervenkr 108:407–431
3. Bickford RG, Butt HR (1955) Hepatic coma: the electroencephalographic pattern. J Clin Invest 34:790–799
4. Bok ST, Shade JP (1957) Hypothermia and cerebral activity. Acta Physiol Pharmacol Neerl 6:775–794
5. Boudin G, Lauras A, Labet R (1961) Les encéphalopathies des hypoglyciémies spontanées. Revue analytique de 37 cas à propos d'une observation personelle. World Neurol 2:849–861
6. Brown-Sequard ME (1857) Recherches experimentales sur les proprietés et les usages du sang rouge et du sang noire. Compt Rend Acad Sci 45:562–566
7. Cohn R, Sode J (1971) The EEG in hypercalcemia. Neurology (Minneap) 21:154–161
8. Engel R (1955) Diabetes und Hirnstromaktivität. Nervenarzt 26:247–250
9. Faure J, Loiseau P (1956) Electroencéphalogramme et troubles menstruels. Rev Neurol (Paris) 95:525–530
10. Gibbs FA, Williams D, Gibbs EL (1940) Modification of the cortical frequency spectrum by changes in CO_2, blood sugar and O_2. J Neurophysiol 3:49–58
11. Grønquist YKJ, Seldon TH, Faulconer A Jr (1952) Cerebral anoxia during anaesthesia: prognostic significance of electroencephalographic changes. Ann. Chir Gynaecol Fenniae 41:149–159
12. Heppenstal ME (1944) Relation between the effects of the blood sugar levels and hyperventilation on the electroencephalogram. J Neurol Neurosurg Psychiatry 7:112–118
13. Holmberg G (1953) Electroencephalogram during hypoxia and hyperventilation. Electroencephalogr Clin Neurophysiol 5:371–376
14. Junega I, Flynn RE, Berger RL (1972) The arterial, venous pressures and the electroencephalogram during open heart surgery. Acta Neurol Scand 48:163–168
15. Kollmannsberger A (1970) The EEG in liver disease. Electroencephalogr Clin Neurophysiol 29:214
16. Kornmüller AE, Palme F, Strughold H (1941) Über Veränderungen der Gehirnströme im akuten Sauerstoffmangel. Luftfahrtmedizin 5:161–183
17. Krankenhagen B, Penin H, Zeh W (1970) Prä- und postoperative EEG-Untersuchungen bei Patienten mit Cushing-Syndrom. EEG EMG 1:14–19

18. Krump JE (1956) Die klinische Bedeutung des Elektroenzephalogramms bei Vergiftungen, Endotoxikosen und Endokrinopathien. Dtsch Int Tag Leipzig: symposium, 1955. VEB-Verlag Volk und Gesundheit, Berlin, p 133–136
19. Kubicki S, Trede M, Just O (1960) Die Bedeutung des EEG bei Herzoperationen in Hypothermie und bei extrakorporaler Zirkulation. Anaesthesist 9:119–123
20. Kugler J (1972) Zerebrale ischämische Krisen – von der aktivierten partiellen Krise zur spontanen Synkope. EEG EMG 3:109–120
21. Lenard HG, Bell EF (1973) Bioelectric brain development in hypothyroidism. A quantitative analysis with EEG power spectra. Electroencephalogr Clin Neurophysiol 35:545–549
22. Martin JT, Faulconer A, Bickford RG (1959) Electroencephalography in anesthesiology. Anesthesiology 20:359–376
23. Niedermeyer E (1972) The EEG in cardiac diseases. In: Remond A (ed) Handbook of electroencephalography and clinical neurophysiology, vol 14A, Elsevier, Amsterdam, pp 65–67
24. Nieman EA (1959) The electroencephalogram in myxoedema coma. Clinical and electroencephalographic study of three cases. Br Med J I:1204–1208
25. Örley J, Tomka I, Csaki P, Nagy M (1973) Electroencephalographic studies in perimenarchial girls. Acta Paediatr Acad Sci Hung 14:69–79
26. Penin H (1967) Über den diagnostischen Wert des Hirnstrombildes bei der hepato-portalen Enzephalopathie. Zugleich ein klinisch-statistischer Beitrag zur Frage neurologischer und psychischer Veränderungen bei Leberzirrhosen und porto-cavaler Anastomosenoperation. Fortschr. Neurol Psychiatr 35:174–234
27. Petrov IR (1939) Changes in function of central nervous system in anoxemia of various forms. Klin Med (Mosk) 17:6–24
28. Pitot M, Gastaut H (1953) Modifications électroencéphalographiques au cours du cycle œstrol. Rev Neurol (Paris) 89:427–430
29. Reichenmiller HE, Dürr F, Reinhard U (1970) Schlafuntersuchungen bei urämischen Patienten. Verh Dtsch Ges Inn Med 76:1036–1040
30. Roth B, Nevsimal O (1964) EEG-study of tetany and spasmophilia. Electroencephalogr Clin Neurophysiol 17:36–45
31. Silverman D (1962) Some observations on the EEG in hepatic coma. Electroencephalogr Clin Neurophysiol 14:53–59
32. Spatz R, Kollmannsberger A, Holzer P (1973) Elektroenzephalographische Veränderungen bei Tolbutamid-induzierter Hypoglykämie. EEG EMG 4:131–137
33. Spunda C (1959) Über den Effekt der Beatmung mit verschiedenen Gasen auf das normale und abnormale EEG. Wien Klin Wochenschr 71:513–519
34. Swash M (1972) Electroencephalographic criteria of hypocalcemia and hypercalcemia. Arch Neurol 26:218–228
35. Trede M, Kubicki S, Just O (1959) Über EEG-Beobachtungen bei Herzoperationen mit extrakorporalem Kreislauf. Anaesthesist 8:76–82
36. Tyler HR, Clark DB (1957) Incidence of neurological complications in congenital heart disease. Arch Neurol Psychiatr 77:483–489
37. Vague J, Gastaut H, Codaccioni JL (1957) L'electroencéphalographie des maladies thyroidiennes. Ann Endocrinol (Paris) 18:996–1008
38. Weinberger LM, Gibbon MH, Gibbon JH Jr (1940) Temporary arrest of circulation to central nervous system; physiological effects. Arch Neurol Psychiatr 43:615–634
39. Weinberger LM, Gibbon MH, Gibbon JH Jr (1940) Temporary arrest of circulation to central nervous system; pathological effects. Arch Neurol Psychiatr 43:961–986
40. Wilson WP, Sieker HO (1958) The study of the factors responsible for changes in the electroencephalogram on chronic pulmonary insufficiency. Electroencephalogr Clin Neurophysiol 10:89–96
41. Zander-Olsen P, Støier M, Siersbaek-Nielsen K, Mølholm-Hansen J, Schiøler M (1972) Electroencephalographic findings in hyperthyroidism. Electroencephalogr Clin Neurophysiol 32:171–177
42. Zysno E, Dürr F, Reichenmiller HE, Nieth H (1967) EEG–Untersuchungen bei urämischen Enzephalopathien unter intermittierender Peritonealdialyse. Verh Dtsch Ges Inn Med 72:227–299

IX. Special Considerations in Old Patients

General life expectancy has constantly risen in Western societies through improved living conditions and medical care. Today, it ranges between 75 and 80 years, with, consequently, a great number of geriatric patients to be managed anesthesiologically. In older patients, decreased functions affecting individual or all organs to various degrees have to be taken into account. In contrast to younger patients, these decreased functions lead to a reduced state of health, which is reflected in the risk-group classification of the American Society of Anesthesiologists (ASA) [2] in the score of two to three.

EEG Findings

Our Findings (250 patients over 70 years of age) (Figs. 1–3). In patients over 70 years of age with no previous cerebral abnormality, we recorded a normal EEG in 54% of cases and an altered EEG in 46% (see Chap. A IV) (Fig. 1). In old patients with a normal initial EEG, electric functional changes as a result of anesthesia were similar to those in younger patients (Fig. 2). In patients with abnormal initial EEG patterns, responses of the frequency to anesthesia deviated from normal; here, the extent of the initial difference is important. With a general EEG deceleration, specific pharmaceutical reactions and stages of anesthesia can still be determined by the changes produced. With an irregular initial EEG, usually ranging from the delta to the mid-beta range, only one single reaction is provoked by primarily different drugs, i. e., a limitation of the frequency range in the beta band. Although this uniform reaction allows a certain continuous check of anesthesia with continuous monitoring, it exclude a more precise evaluation of individual stages of anesthesia (Fig. 3).

Review of Other Findings. No direct EEG studies have been performed on geriatric patients under general anesthesia.

Clinical Evaluation. Anesthesiological experience has shown that older patients react more sensitively to anesthetics than would normally be expected [1]. Therefore, it is necessary to use drugs for induction at half the recommended dose and to inject them slowly. This "sensitivity of anesthesia" mainly concerns the circulatory reactions visible and measurable in routine anesthesia; these reactions are more pronounced in older patients and can-

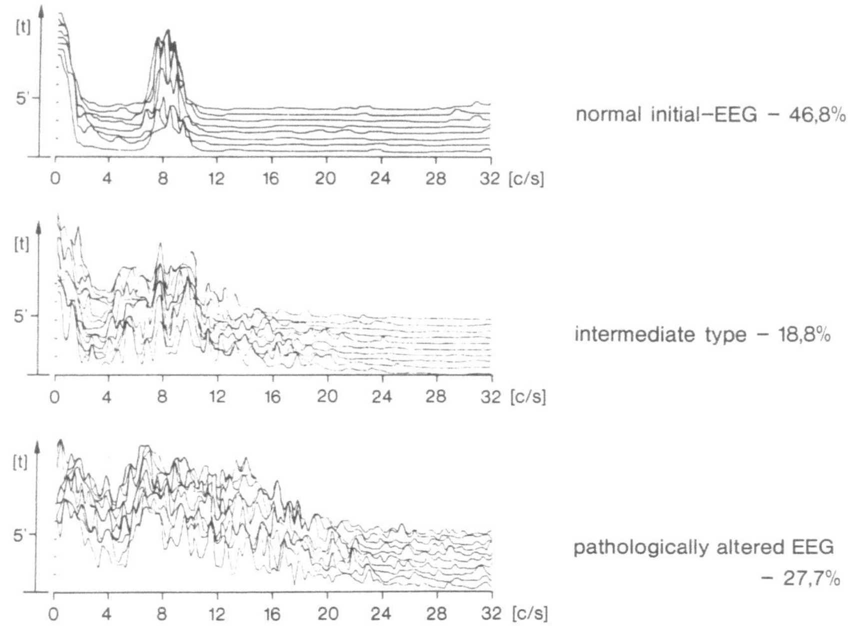

Fig. 1. Distribution of normal, slightly pathological, and pathological resting EEG in geriatric (> 70 years of age) patients undergoing surgery ($n = 250$)

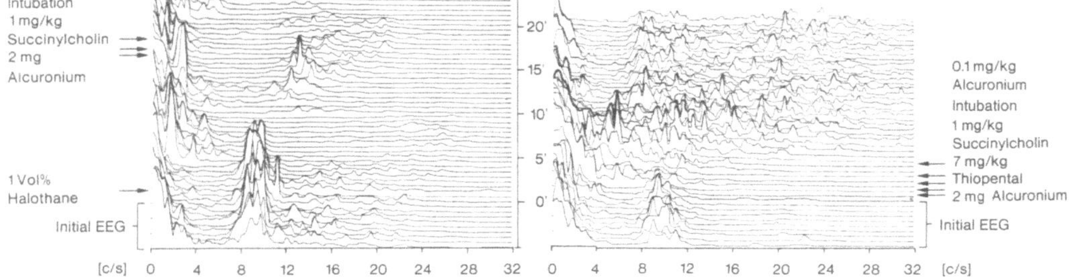

Fig. 2. Cerebral reaction to induction with halothane **a** and thiopental **b** in geriatric patients with a normal resting EEG. In both kinds of anesthesia, there are the typical frequency patterns characteristic for each drug at induction.
Recording conditions: TC 0.3 s; filter, 70 Hz; calibration, 50 μV = 7 mm; paper speed, 30 mm/s; FFT, 30-s periods

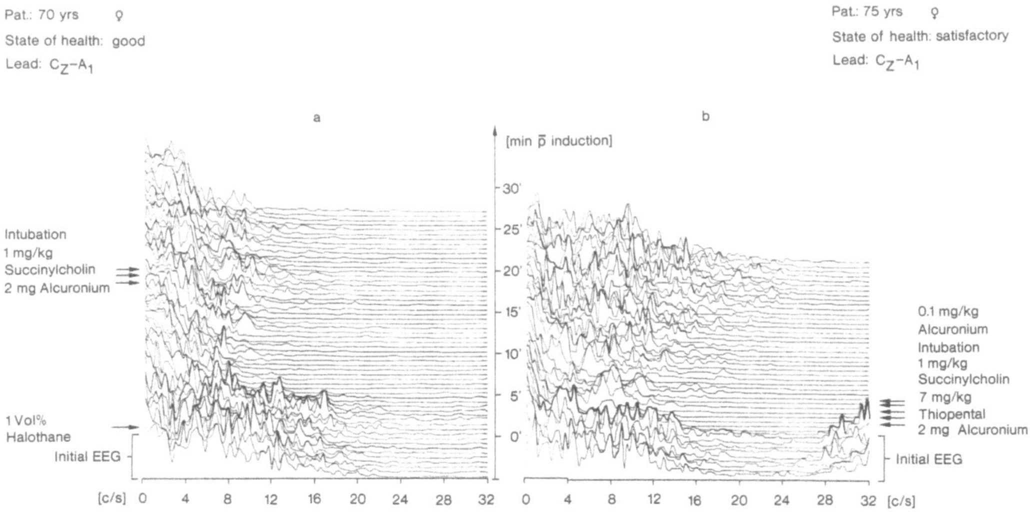

Fig. 3. Cerebral reaction during halothane **a** or thiopental **b** induction in the pathological initial EEG of geriatric patients. The figures demonstrate the absence of drug-specific reactions in the EEG, which are characterized by a stereotyped limitation of the upper frequency of an irregular initial EEG. A definite classification of the anesthetic depth is no longer possible. Recording conditions: TC, 0.3 s; filter, 70 Hz; calibration, 50 μV = 7 mm; paper speed, 30 mm/s; FFT, 30-s periods

not be compensated for so well, thus requiring counter measures from the anesthesiologist. The reaction of the all other organs in routine anesthesia is not immediately evident to the anesthesiologist and may actually be obscured. The more marked reaction of the brain cells on weight-related amounts of analgetics, tranquilizers, or induction anesthetics – with the standard rate of injection – has been discussed in Chaps. B I and B III. In older patients, deeper sedative or anesthetic stages are produced than in younger patients given equivalent drug doses; moreover, the effect lasts considerably longer.

Another clinical experience is the observation that, frequently, old patients are especially difficult to anesthetize. In particular, when continuing anesthesia a higher dose than previously estimated is often required, leading postoperatively to considerable hangovers. Thus, in the study concerning neuroleptanalgesia, discussed in Chap. B IV, there is a higher drug consumption in geriatric patients. Explanations for such variations have thus far not been found in the anticipated cerebral response.

Interpretation of Findings in Old Patients

Despite the limitations of the information to be derived from an EEG during anesthesia in some cases, as described above, EEG monitoring appears to be a useful procedure in old patients: Cerebral emergencies here are also recognized by a further decrease in frequency, but mainly – at a somewhat later stage – by a decrease in amplitude. Surprisingly, according to our observations in geriatric patients, cerebral deficiency symptoms, e.g., due to hypotension occur later than in younger subjects, but lead more rapidly to sustained or irreversible changes.

References

Papers

1. Castleden CM, George CF, Marcer D, Hallet C (1977) Increased sensitivity to nitrazepam in old age. Br Med J 1:10–12
2. Dripps RD, Eckenhoff JE, Vandom LD (1967) Physical status as it relates to risk. In: Dripps RD, Eckenhoff JE, Vandom LD (eds) Introduction to anesthesia, 3rd edn. Saunders, Philadelphia, pp 15–18

X. Special Considerations in Patients with Seizure Disorders

CONTENTS

Among the patients undergoing anesthesia for surgery there are cases involving cerebral convulsions as concomitant findings or as a symptom of the disease under treatment. These cases include epilepsy, which today can usually be treated with medication (approx. 0.5–2% of the standard population [9, 21]), latent epilepsy (approx. 0.6% of the normal population [12]), and seizures as sequelae of cerebral injuries or as a symptom of growing and displacing leasions in the brain. Special aspects of anesthesiological care need to be considered in the preparation and premedication, during the phase of anesthesia, and in the recovery period.

Preparation and Premedication

The medical history often reveals the kind, degree, and frequency of the seizures and the extent to which they can be controlled using drugs. Preoperative EEG findings should be available despite their limited diagnostic value. Depending on the kind of seizures, the incidence of positive EEG findings ranges between 15% and 96% [5]; grand mal epilepsies, which are the most frequent, show normal EEG findings in 36% of cases (Fig. 1) [5]. Although the EEG cannot accurately depict seizure-disorders, typical generalized or focal spikes, sharp waves, spike-wave, and sharp slow-wave complexes can provide diagnostic indications unless the seizures disorders is already known from the anamnesis (Fig. 2). The incidence and frequency of such graphoelements indicate the actual cerebral function, which may be improved by a modification of the preoperative therapy. With good clinical control, however, no further EEG normalization is required. Anesthesiological measures as well as changes in the psychological state may provoke seizures perioperatively through various mechanisms. The antiepileptic

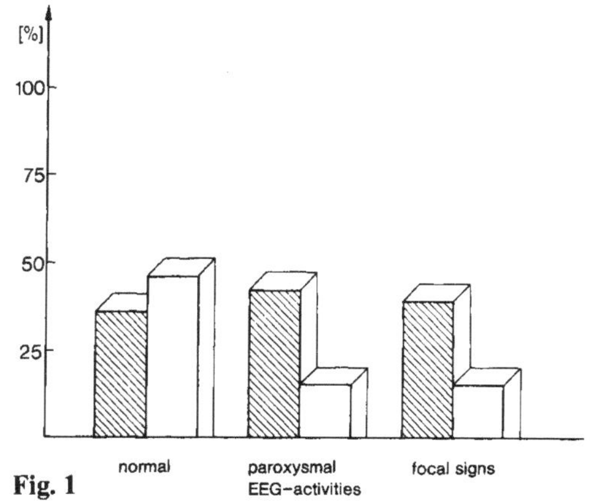

Fig. 1

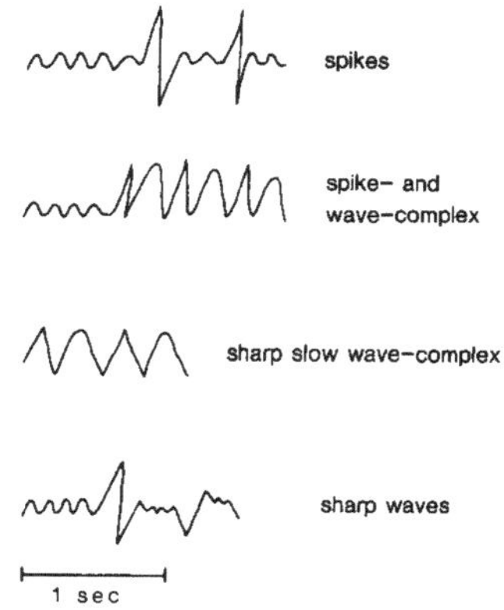

Fig. 2

Fig. 1. Percentage of abnormal EEG findings with typical ▨ and abortive ☐ grandmal seizures (after DEGEN [5]

Fig. 2. Typical patterns of paroxysmal EEG activitiy in convulsive disorders

medication should be continued until immediately prior to anesthesia and not be discontinued until 6 h before the induction of anesthesia, with no food being consumed by the patient. In the selection of preoperative sedatives, substances having an anticonvulsive component should be favored. Diazepam is usually taken to be the drug of choice [20]. The combination of pethidine and promethazine is not optimal for premedication in this particular context because excitation phenomena can occur after pethidine application and neuroleptics, e.g., promethazine, should not be used in the treatment of epileptics due to their possible seizure-triggering effects [3].

Induction and Maintenance of Anesthesia

An awareness of the possible excitatory effects of anesthetics and similar agents is essential for the induction of general anesthesia in patients suffering from seizures. WINTERS et al. [23–25] have classified the anesthetics currently used into cerebrally excitant and depressive substances. Accordingly, excitant anesthetics, e.g., ketamine and enflurane, should frequently cause seizures, whereas depressive anesthetics, e.g., barbiturates and halothane should generally remain inert. This classification has not been confirmed in routine practice. According to KUGLER and DOENICKE [12], in the reference material between 1973 and 1977 the rate of seizures in each of the abovementioned groups was similar. Other reports show further contradictions: On the basis of EEG studies [6], etomidate, e.g., is identified as a substance potentially dangerous for an epileptic – clinically, however, only myoclonias have been reported, no epileptic fits. Animal experiments and clinical applications indicate a strong anticonvulsive effect of etomidate [18, 19]. With ketamine and ethrane, on the one hand, EEG changes with the occasional occurrence of convulsive potentials are described; on the other hand, both substances have been recommended for epileptics [1, 4, 8, 17].

Regional and local anesthesia is generally considered to be contraindicated in diseases of the CNS [2, 13, 14]. More recent studies, however, demonstrate an anticonvulsive effect of local anesthetics at low blood levels [7, 10, 15, 22]. Spinal and peridural anesthesias have successfully been performed on patients [11] seizures; the cerebral state of the alert patient can be checked continuously, and the oral medication may be resumed immediately postoperation. The question of whether changes in the CSF pressure in spinal anesthesia or accidental perforations of the dura in peridural anesthesia can provoke seizures is still open. With current uncertainty as to the choice of anesthesia for patients suffering from seizures, the following risks from anesthesia must be considered. Physiological sleep is one of the strongest means of provoking certain kinds of epileptic convulsions; increased susceptibility to convulsions occurs on the threshold of wakefulness, not in the deep stages of anesthesia [12]. Thus, the danger of provoking a seizure is greatest during the induction and the waking phases of anesthesia. Drugs providing a quick transition from alertness to the desired depth of anesthesia and a quick resumption of full consciousness in the recovery phase are preferred. No particular medication can be recommended here; the anesthesiologist makes his choice in the light of his own experience.

Fluctuations in the serum level of antiepileptics through preoperative starvation of surgical procedures are in the order of 15–30% [16]. Such fluctuations are also possible within the normal daily profile, implying no special proneness to seizures. In general, the intraoperative application of anticonvulsants can, therefore, be dismissed [16]. Only with surgery of very long duration might this be done, whereby interaction with the hypnoanalgetics administered for anesthesia (synergistic and/or potentiating effects) has to be considered. An intraoperative EEG check by means of a conventional recording is expedient; spectral analytic methods, however, in general do not allow any pattern diagnosis to be made; they demonstrate the cerebral background activity in the EEG, but only reveal paroxysmal activity to a limited degree.

Postoperative Phase

The rapidity at which the individual stages of anesthesia are passed through during induction can be varied within certain limits by the amount and speed of injection of the hypnoanalgetic. In contrast, the recovery phase is determined by elimination of the narcotic, which cannot be influenced by the anesthesiologist. Rather, it is determined by the individual situation of the patient and by physical/chemical properties of the anesthetics. Since there are not other possibilities of influencing this phase is recommended the patient be closely monitored so that in the event of a seizure immediate steps can be taken. The observation period should cover at least 6 h, because definite normalization of the anesthesia-induced cerebral changes cannot be expected any earlier. Postoperative resumption of the anticonvulsive therapy must be attempted as soon as possible. If postoperative starvation becomes necessary, any application has to be given parenterally; the scheme should be devised by the neurologist prior to surgery.

Final Evaluation

Anesthesia with the drugs available today is possible in the epileptic patient and can be kept within a reasonable range of risk through special care in the postoperative period. Regional and local anesthesia also seem relatively harmless; however, they are still today contraindicated in patients suffering from seizures. In evaluating the proneness to seizures, EEG monitoring during anesthesia is at present only useful in the conventional form, because the graphoelements in question cannot be determined using spectral analytic methods.

References

Papers

1. Arfel G, Weiss J (1962) Paroxysmal EEG-patterns and convulsive states in the course of general anesthesia. Electroencephalogr Clin Neurophysiol 14:788
2. Bergmann H (1982) Die Spiralanästhesie. In: Benzer H, Frey R, Hügin W, Mayrhofer O (eds) Anaesthesiologie, Intensivmedizin und Reanimatologie. Springer, Berlin Heidelberg New York, pp 482–491
3. Busse O, Künkel H (1978) Status epilepticus – Klinik und Therapie. Dtsch Ärzteblatt 75:1987–1992
4. Corssen G, Little SA, Tavakoli M (1974) Ketamine and epilepsy. Anesth Analg (Cleve) 53:319–335
5. Degen R (1980) EEG-Befunde bei zerebralen Krampfanfällen. In: Opitz A, Degen R (eds) Anästhesie bei zerebralen Krampfanfällen und Intensivtherapie des Status epilepticus. Perimed, Erlangen, p 133–143
6. Degen R, Oberwetter W-D, Degen H-E, Koch D (1980) Etomidat zur Narkoseeinleitung bei Epileptikern. In: Opitz A, Degen R (eds) Anästhesie bei zerebralen Krampfanfällen und Intensivtherapie des Status epilepticus. Perimed, Erlangen, p 87–118
7. Demetrescu M, Julien RM (1974) Local anesthesia and experimental epilepsy. Epilepsia 15:235–248
8. Faure C, Fihey A, Lupold M (1974) L'éncephalographie gazeuse sous anesthésie generale au chlorhydrate de ketamine chez l'enfants. J Radiol Electrol Med Nucl 56:717–720
9. Janz D (1967) Typen der Epilepsie-Verläufe. Therapiewoche 3:67–70

10. Julien RM (1973) Lidocaine in experimental epilepsy: correlation of anticonvulsant effect with blood concentrations. Electroencephalogr Clin Neurophysiol 34:639–645
11. Koch D, Degen H-E, Oberwetter W-D, Degen R, Opitz A (1980) Regionalanästhesie bei Patienten mit zerebralen Krampfleiden. In: Opitz A, Degen R (eds) Anästhesie bei zerebralen Krampfanfällen und Intensivtherapie des Status epilepticus. Perimed, Erlangen, pp 127–132
12. Kugler J, Doenicke A (1980) Anfälle bei Narkose? In: Opitz A, Degen R (eds) Anästhesie bei zerebralen Krampfanfällen und Intensivtherapie des Status epilepticus. Perimed, Erlangen, pp 133–143
13. Lee JA, Atkinson RS (1978) Synopsis der Anästhesie. Fischer, Stuttgart New York, p 281–412
14. Nemes C, Niemer M, Noak G (1979) Datenbuch der Anästhesiologie. Fischer, Stuttgart, p 136
15. Nolte H (1980) Die Wirkung der Lokalanästhetika auf das zentrale Nervensystem. In: Opitz A, Degen R (eds) Anästhesie bei zerebralen Krampfanfällen und Intensivtherapie des Status epilepticus. Perimed, Erlangen, pp 123–126
16. Oberwetter WD (1980) Einfluß der Infusionstherapie auf das Verhalten der Serumspiegel antiepileptischer Medikamente. In: Opitz A, Degen R (eds) Anästhesie bei zerebralen Krampfanfällen und Intensivtherapie des Status epilepticus. Perimed, Erlangen, pp 57–64
17. Opitz A, Brecht S, Stenzel E (1977) Enfluran-Anästhesien bei Epileptikern. Anaesthesist 26:329–332
18. van der Starre P (1980) Etomidat als schnellwirkende antiepileptische Substanz. In: Opitz A, Degen R (eds) Anästhesie bei zerebralen Krampfanfällen und Intensivtherapie des Status epilepticus. Perimed, Erlangen, pp 205–215
19. Wauquier A, Ashton D, Clincke G, Niemeggers CJE, Janssen PAJ (1980) Etomidate: A non-barbiturate hypnotic; anticonvulsant, anti-toxic and brain protective actions in animals. In: Opitz A, Degen R (eds) Anästhesie bei zerebralen Krampfanfällen und Intensivtherapie des Status epilepticus. Perimed, Erlangen, pp 183–203
20. Weber R, Koch D (1980) Prämedikation bei Epileptikern. In: Opitz A, Degen R (eds) Anästhesie bei zerebralen Krampfanfällen und Intensivtherapie des Status epilepticus. Perimed, Erlangen, pp 43–44
21. Went LN (1979) Epilepsie und genetische Beratung. In: Doose H, Dam M, Gross-Selbeck G, Meinardi H (eds) Epilepsie 1979. Tagung der Deutschen, Dänischen, Holländischen und Schweizerischen Sektionen der internationalen Liga gegen Epilepsie. Regionaler Kongreß von Epilepsy International. Thieme, Stuttgart, pp 109–113
22. Wikinski JA, Usubiaga JE, Morales RL, Torriesi A, Usubiaga LE (1970) Mechanism of convulsions elicited by local anesthetic agents. Anesth Analg Curr Res 49:504–510
23. Winters WD (1972) Epilepsy of anesthesia with Ketamine. Anesthesiology 36:309–312
24. Winters WD, Mori K, Spooner CE, Bauer RO (1967) The neurophysiology of anesthesia. Anesthesiology 28:65–80
25. Winters WD, Wallbach MB (1970) Drug induced states of CNS excitation. A theory of hallucinosis, psychosomimetic drugs. In: Efron DH (ed) Raven, New York, pp 193–214

C. Electroencephalography as a Method of Anesthesiological Monitoring

I. Electroencephalographic Monitoring During Anesthesia

CONTENTS

Information about actual cerebral function through continuous EEG monitoring facilitates a precise, individually controlled depth of anesthesia. It also indicates dangers or impairments to the cerebrum due to surgical or anesthesiological distress. As an early, and frequently the only, warning symptom, it can enable either a diagnosis to be made in good time, with the possible avoidance of cerebral oxygen deficiency, or it can be of help in the carrying out of an appropriate therapy. EEG monitoring, therefore, has advantages both for the patient with a limited cerebral compensatory reserve and for patients undergoing extensive surgery. Knowledge of the drug-induced alterations is again required for safely evaluating sudden EEG changes. Cumulative effects of narcotics, inevitable during long surgery, involve additional deviations from the normal EEG pattern, which have to be taken into account.

Barbiturate-Induced Inhalation Anesthesia

Thiopental-Halothane

Today, thiopental is the leading anesthetic for induction in inhalation anesthesia. The barbiturate dose in induction bridges the relatively long time required for inhalation anesthetics to take effect, this period being usually unpleasant for the patient. Anesthesia is then continued as inhalation anesthesia, using curarelike muscle relaxants.

EEG Findings

Our Findings (46 patients, 20–80 years of age, dose 5 mg/kg thiopental, 1 vol.% halothane, carrier gas N_2O/O_2 3 : 1) (Table 1; Fig. 1). With barbiturate induction, the initial phase of halothane, which is usually clearly visible in the EEG (excitation in the form of fast beta frequencies) is superimposed and characterized by typical signs of barbiturates. An irregular EEG (delta/theta/alpha/beta) results within the first 2–3 min. In the following 15–20 min of deepening of anesthesia, the EEG demonstrates a dominance of the slow frequency portions (delta/theta), with alpha/beta components still in evidence.

Table 1. EEG pattern during induction with halothane-nitrous oxide-oxygen anesthesia, preceded by barbiturates ($n = 46$)

Frequency range	Nature of EEG power change[a]		Degree of change	Drug dose
	During barbiturate phase	During halothane phase		
Alpha	Typical barbiturate effect in 100%	76% of cases	< 22 µV	Thiopental 5 mg/kg
Beta	2.25 ± 1.2 min sustained, then:			Succinylcholine 1 mg/kg for intubation
Delta	subsequent transition to halothane	+ alpha/beta 76%	< 24 µV	Halothane 1 vol. %
Theta	phase	dominant 24%		Respiration with
Irregular		0	0	N_2O/O_2 3 : 1

[a] ↑ increase, 0 absence

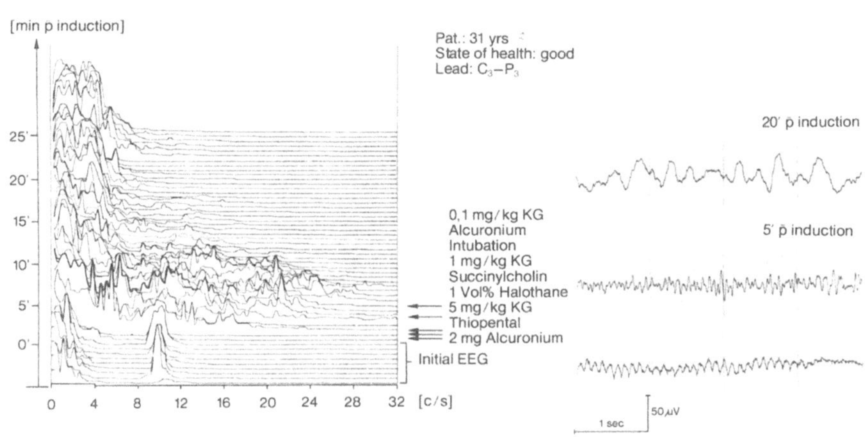

Fig. 1. EEG spectral analysis and parts of the conventional EEG during induction of anesthesia with a combined thiopental-halothane mixture. After injection of thiopental, there are typical barbiturate-induced EEG changes lasting about 7 min. Subsequently, there is the immediate pattern of a deep halothane anesthesia with high-voltage delta/theta activity.
Recording conditions: TC, 0.3 s; filter, 70 Hz; calibration, 50 µV = 7 mm; paper speed, 30 mm/s; FFT, 30-s periods

Findings During Anesthesia

Our Findings (16 patients, 20–80 years of age, halothane N_2O/O_2 and muscle relaxants as needed) (Tables 2, 3; Fig. 2). The EEG findings in continuous intraoperative recording show deep (delta/theta) or sufficiently deep (delta/theta/alpha) anesthesia in 68.8% of patients. In 12.5% of patients, there is a slight activation of all four frequency bands and in 18.7% there are wide changes in the stage of anesthetic depth (delta/theta/alpha/beta).

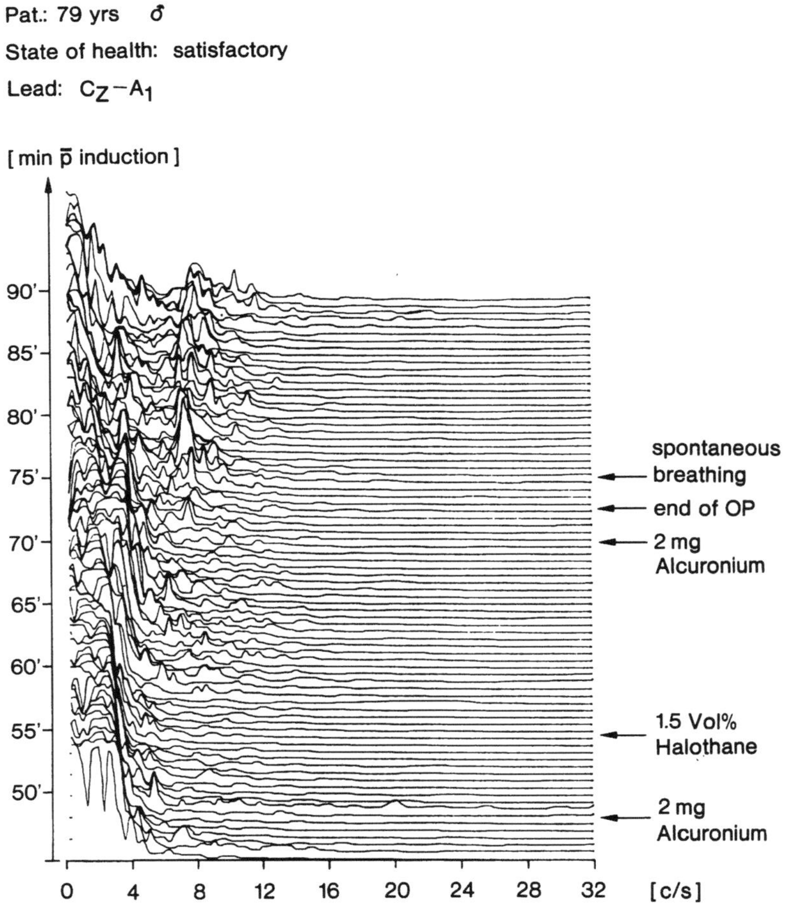

Fig. 2. EEG spectral analysis during barbiturate-induced halothane anesthesia. The EEG demonstrates the desired course of anesthesia – slow high-voltage frequencies (0.5–4 Hz) during intra-abdominal manipulation, which accelerate toward the end of the surgery and at the time of extubation are replaced by slow alpha activitiy. With regard to the course of the operation, the depth of anesthesia was appropriate and is well documented in the EEG.
Recording conditions: TC, 0.3 s; filter, 70 Hz; calibration, 50 μV = 7 mm; paper speed, 30 mm/s; FFT, 30-s periods

In some cases, particular kinds of cerebral distress (transient decreased blood pressure; transition from machine respiration to spontaneous breathing) cause acute impairment of electric brain function (altogether, in 25% of patients). Such cerebral distress is evident as a sudden frequency deceleration, possibly as a decrease in voltage in the EEG, and indicates a deficient supply of oxygen to the brain. A comparison between the younger and older age-groups shows that in older patients deeper stages of anesthesia are achieved with doses of anesthetics that are determined by clinical fac-

Table 2. Anesthetic requirements and patient data with thiopental-halothane anesthesia (Mean ± standard deviation ($\bar{x} \pm s\bar{x}$)

Drug	Total ($n = 16$)		< 50 years ($n = 8$)		> 70 years ($n = 8$)	
Thiopental (mg/kg)	5.53	± 1.14	5.87	± 0.92	5.20	± 1.30
Succinylcholine (mg/kg)	1.21	± 0.38	1.35	±0.21	1.07	± 0.47
Alcuronium (mg/kg)	16.3608×10^{-4} $\pm 7.4555 \times 10^{-4}$		16.2296×10^{-4} $\pm 9.7995 \times 10^{-4}$		16.4923×10^{-4} $\pm 4.8013 \times 10^{-4}$	
Halothane (l/kg/min)	10.7377×10^{-4} $\pm 2.3552 \times 10^{-4}$		10.1824×10^{-4} $\pm 1.5638 \times 10^{-4}$		11.2929×10^{-4} $\pm 2.9557 \times 10^{-4}$	
Postsurgical artificial respiration in ICU (*n*, percent)	3	18.75	0	0	3	37.5
State of health (1–4)[a]	2.06	± 0.77	1.75	± 0.93	2.4	± 0.44
Duration of anesthesia (min)	164.7	±59.7	175.0	±66.3	154.4	±54.7

Abbreviation: ICU, intensive care unit

[a] According to American Association of Anesthesiologists (ASA) recommendation

Table 3. Intraoperative EEG findings with barbiturate-induced halothane anesthesia

Electroencephalographically defined depth of anesthesia	Total ($n = 16$)		< 50 years ($n = 8$)		> 70 years ($n = 8$)	
	n	Percent	*n*	Percent	*n*	Percent
Deep anesthesia delta/theta	7	43.75	1	12.5	6	75
Sufficient anesthesia delta/theta/alpha	4	25	3	37.5	1	12.5
Light anesthesia alpha/beta or theta	2	12.5	2	25	0	0
Widely fluctuating anesthesia delta/theta/alpha/theta	3	18.75	2	25	1	12.5
Impaired cerebral function due to exogenous factors	4	25	3	37.5	1	12.5

tors. On the one hand, this necessitates postoperative machine respiration in intensive care units; conversely, it offers a certain protection against adverse effects of exogenous factors due to the decrease in metabolism. Nevertheless, any interfering factors can still be determined from the EEG.

Clinical Evaluation (Table 2). The consumption of anesthetics and relaxants with clinically controlled doses corresponds to a standard moderately long surgical procedure of approximately 2½ h. The state of health of all patients, with an American Society of Anesthesiologists (ASA) score of 2.1, is considered moderate. Clinically, the following differences are seen between the two selected age-groups. The health condition of the geriatric patients is markedly poorer, with an ASA score of 2.4 as compared with 1.8 in the younger patients. Anesthesia in the young patients is about 20 min longer. The higher consumption of halothane and relaxants in the group of the older patients is significant. Postanesthetic respiration because of narcotic hangover is necessary in over one-third of cases in the age-group of 70 years and older.

Interpretation of EEG Findings and Clinical Observations. As described in the previous chapters, the anesthetic stages in combination anesthesia with thiopental/halothane can be distinguished and accurately determined. Intraoperative distress, such as extreme lowering of blood pressure or respiratory hypoxia or anoxia, soon becomes apparent in the EEG changes, possibly even prior to the occurrence of cardiovascular signs of an oxygen deficit. The clinical observation of higher consumption of anesthetics in older patients is reflected in the EEG in deeper stages of anesthesia. Thus, the EEG in thiopental/halothane anesthesia allows a determination of both the depth of anesthesia as well as the extent of cerebral effects by exogenous influences.

Thiopental-Enflurane

Various anesthesiological aspects (rapid pharmacological action; minimal sensitization to catecholamines; minimal influence on uterine tone) have led to a gradually increasing application of the combination of thiopental and enflurane. The lower metabolic rate of enflurane, compared with halothane, is thought to produce less toxicity, although this is debatable, particularly with regard to the kidney. The dose of enflurane is approximately double that of halothane due to the lower narcotic potential.

EEG Findings During Anesthesia

Our Findings (17 patients, 20–75 years of age, enflurane-N_2O/O_2 and muscle relaxants as required) (Tables 4, 5; Figs. 3, 7b). The 17 thiopental-

Table 4. Anesthetic requirements and patient data with thiopental-enflurane anesthesia ($\bar{x} \pm s\bar{x}$)

Drug	Total ($n=17$)		< 50 years ($n=9$)		> 70 years ($n=8$)	
Thiopental (mg/kg)	4.81	± 1.38	5.04	± 1.49	4.56	± 1.29
Succinylcholine (mg/kg)	1.17	± 0.20	1.20	± 0.25	1.13	± 0.12
Alcuronium (mg/kg/min)	16.5149×10^{-4} $\pm 6.2226 \times 10^{-4}$		16.7258×10^{-4} $\pm 5.5803 \times 10^{-4}$		16.2500 ±7.3241	
Enflurane (l/kg/min)	10.3007×10^{-4} $\pm 2.9422 \times 10^{-4}$		11.4565 ±2.4685		9.0005 ±3.0287	
Postsurgical artificial respiration in ICU (n, percent)	3	17.64	1	11.11	2	25.00
State of health (1–4)	1.72	± 0.77	1.33	± 0.5	2.13	± 0.83
Duration of anesthesia (min)	122	±42	133	±41	111	±42

Table 5. Intraoperative EEG findings with barbiturate-induced enflurane anesthesia

Electroencephalographically defined depth of anesthesia	Total ($n=17$)		< 50 years ($n=9$)		> 70 years ($n=8$)	
	n	Percent	n	Percent	n	Percent
Deep anesthesia delta/theta	2	11.75	1	11.11	1	12.50
Sufficient anesthesia delta/theta/alpha	8	47.06	5	55.55	3	37.50
Light anesthesia alpha/beta or theta	7	41.18	3	33.33	4	50.00
Impaired cerebral function due to exogenous factors	1	5.88	0	0	1	12.50

enflurane anesthesias demonstrate principally the same changes in the EEG as with the combination of thiopental and halothane. Marked light stages of anesthesia, according to the EEG, are more frequent with enflurane. Alterations in cerebral function through exogenous influences occurred in 1 of 17 patients (5.9%; patient of the older age-group). Both age-groups show approximately the same distribution of narcotic stages. Conventional EEG recordings did not reveal any signs of generalized abnormal potentials in terms of an epileptogenic potency of enflurane.

Clinical Evaluation (Table 4). The amount of enflurane required for the maintenance of anesthesia is not the same as that for thiopental-halothane. The strong hypotensive effect of the substance is the cause of the compara-

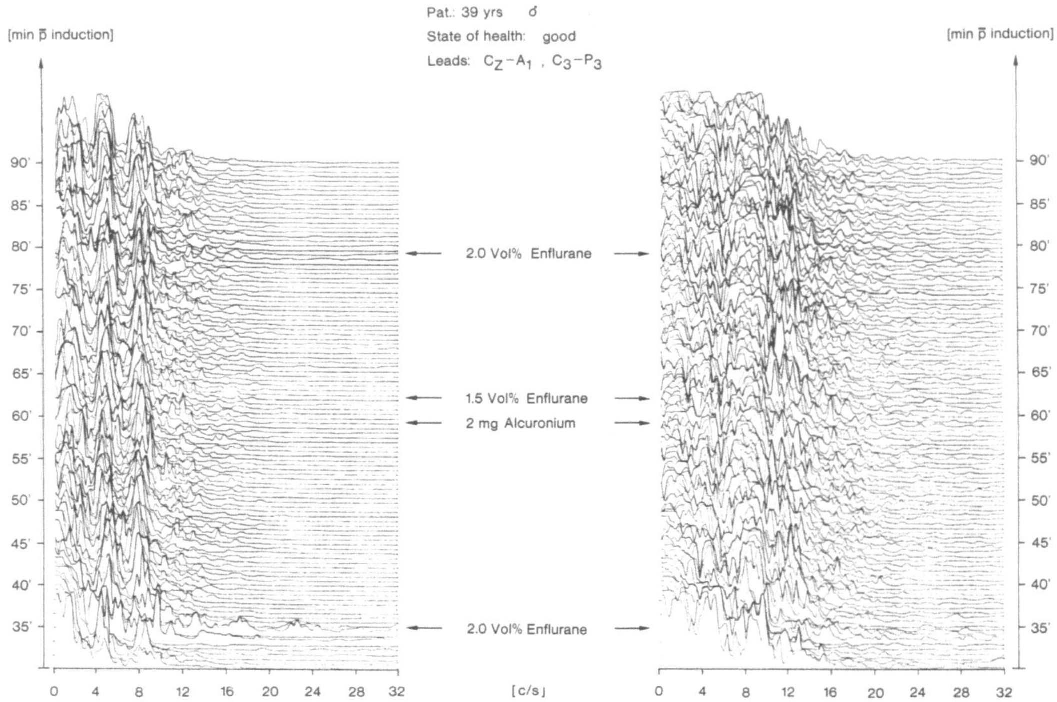

Fig. 3. EEG spectral analysis of barbiturate-induced enflurane inhalation anesthesia. Constant delta/theta/alpha and beta activation (mainly in lead C_Z–A_1) demonstrate the low enflurane dose leading to a just sufficient or light anesthetic depth.
Recording conditions: TC, 0.3 s; filter, 70 Hz; calibration, 50 μV = 7 mm; paper speed, 30 mm/s; FFT, 30-s periods

tively low dosage. In hypovolemic patients, or in patients with a latent or manifest cardiac failure, any higher concentration should be avoided. Comparable enflurane doses were used in both age-groups. The time of surgery in younger patients was also approximately 20 min longer. The state of health shows about the same distribution as for the thiopental/halothane group.

Interpretation of EEG Findings and Clinical Observations. With the combination thiopental/enflurane, the various stages of anesthesia are electroencephalographically distinct. Reductions in the enflurane dose as a result of cardiovascular side effects lead to lighter anesthesia in the EEG. Exogenous hazards, giving rise to a cerebrally deficient supply of oxygen, are manifest immediately in the EEG with this kind of anesthesia.

Barbiturate-Induced Neuroleptanalgesia Thiopental-Fentanyl-Droperidol

The maintenance of anesthesia with analgetic and neuroleptic substances, with the addition of nitrous oxide and muscle relaxants (neuroleptanalgesia), is still one of the gentlest forms of anesthesia for old patients or those in a poor state of health. Minimal toxicity and intraoperative circulatory distress are, however, contrasted by narcotic hangovers, mainly due to the receptor affinity of the opiates. As neuroleptanalgesia (NLA) does not lead to loss of consciousness but to a state of "mineralization," it is occasionally ascribed as exerting insufficient vegetative suppression during surgery. Intraoperative changes in the electric brain pattern are, therefore, of special interest in NLA.

EEG Findings

Our Findings on Maintenance of Anesthesia (16 patients, 20–75 years of age; fentanyl-droperidol, N_2O/O_2 and muscle relaxants as needed) (Tables 6, 7; Figs. 4–6). The EEG changes in NLA during surgery are principally determined by the pattern of the analgetic phase of the NLA, which is extensively described in Chap. B IV. Deep stages of anesthesia were seen only in 19% of cases, mostly in older patients. Adequate light stages of anesthesia were found in 63% of cases; strong fluctuations in depth, partly with surgical stimuli were found in 19% of the patients (25% of the geriatric and 12% of the younger patients). Cerebral alterations due to distress were observed in two cases, i. e., in 25% of the patients.

Clinical Evaluation (Table 6). The mean dose of fentanyl and droperidol for an average duration of surgery of 3 h were 0,95 mg fentanyl and 10 mg droperidol. The dose of fentanyl in geriatric patients was on the whole higher, by 0.3 mg, than in the younger ones. The need for relaxants was generally lower than with barbiturate-induced inhalation anesthesia. Despite the lower need, there was a considerably higher rate of postoperative machine respiration (44%) than with inhalation anesthesia (17–18%). The distribution of the general state of health with thiopental-fentanyl-droperidol was similar to that of the other two groups.

Interpretation of EEG Findings and Clinical Observations. NLA has a special place in EEG monitoring, because it does not pass through the classic EEG stages. Normally, the intraoperative picture of NLA is characterized by its analgetic phase, i. e., by stable, unmodulated, slightly decelerated alpha rhythms, possibly with activities in the delta or theta band. The evaluation of the depth of anesthesia from the EEG requires knowledge of this particular change. The results of our own studies, showing typical EEG patterns in deep anesthesia, even under NLA, suggest a cumulative effect of fentanyl when clinical criteria are used. The stage of insufficient vegetative suppression is indicated by signs of desynchronization. The EEG is, therefore, a suitable aid in NLA for evaluating the quality of anesthesia – furthermore, it may contribute to optimization of the dose.

Table 6. Anesthetic requirements and patient data with thiopental-NLA anesthesia ($\bar{x} \pm s\bar{x}$)

Drugs	Total (n = 16)		< 50 years (n = 8)		> 70 years (n = 8)	
Thiopental (mg/kg)	4.51	± 1.36	4.56	± 1.27	4.47	± 1.54
Succinylcholine (mg/kg)	1.11	± 0.17	1.10	± 0.11	1.10	± 0.22
Alcuronium (mg/kg/min)	12.315 $\times 10^{-4}$ ± 04.500 $\times 10^{-4}$		11.983 $\times 10^{-4}$ ± 04.829 $\times 10^{-4}$		12.703 $\times 10^{-4}$ ± 04.504 $\times 10^{-4}$	
Fentanyl (mg/kg)	9.6135 $\times 10^{-5}$ ± 5.1367 $\times 10^{-5}$		8.0505 $\times 10^{-5}$ ± 1.1020 $\times 10^{-5}$		11.1762 $\times 10^{-5}$ ± 7.2087 $\times 10^{-5}$	
Dehydrobenzperidol (DHB) (mg/kg/min)	100.4756 $\times 10^{-5}$ ± 54.5181 $\times 10^{-5}$		104.1738 $\times 10^{-5}$ ± 68.2994 $\times 10^{-5}$		96.774 $\times 10^{-5}$ ± 40.9022 $\times 10^{-5}$	
Postsurgical artificial respiration on ICU (n, percent)	7	43.75	1	6.25	6	37.5
State of health (1–4)	1.63	± 0.65	1.13	± 0.35	2.13	± 0.44
Duration of anesthesia (min)	199	± 101	206	± 89	193	117

Table 7. Intraoperative EEG findings with barbiturate-induced NLA

Electroencephalographically defined depth of anesthesia	Total (n = 16)		< 50 years (n = 8)		> 70 years (n = 8)	
	n	Percent	n	Percent	n	Percent
Deep anesthesia delta/theta	3	18.75	1	12.5	2	25
Sufficient anesthesia delta/theta/alpha	3	31.5	3	37.5	2	25
Light anesthesia delta/theta alpha/beta – beta	5	31.5	3	37.5	2	25
Large fluctuations in depth of anesthesia delta/theta – alpha/theta	3	18.75	1	12.5	2	25
Impaired cerebral function due to exogenous factors (n, percent)	2	12.5	2	25	0	0

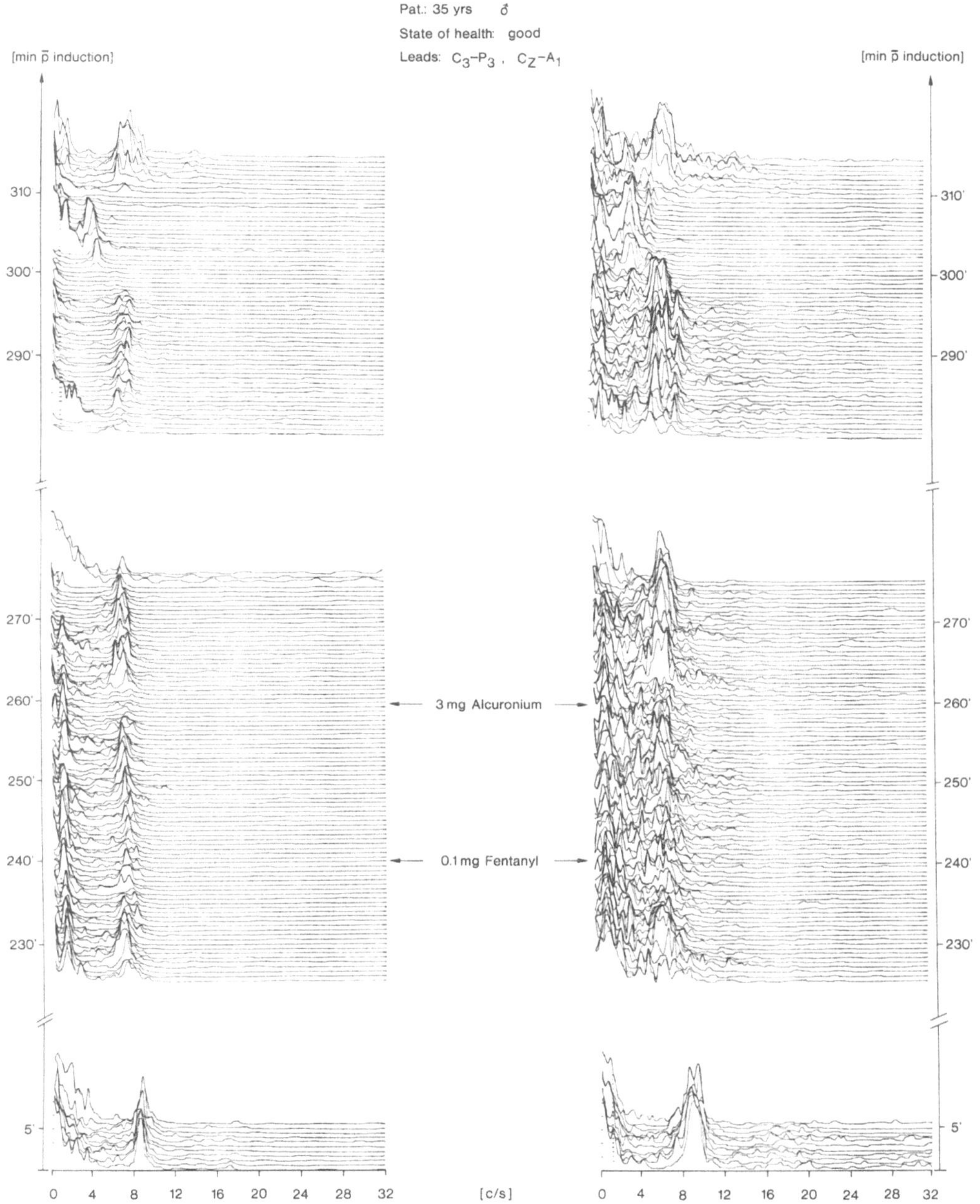

Fig. 4. EEG spectral analysis prior to, during, and in the final stage of barbiturate-induced neuroleptanalgesia. The initial EEG *top* is characterized by a normal alpha rhythm with a dominant frequency of 8 Hz and some beta activity (track C_3-P_3), the latter being considered an effect of premedication. In the steady state of anesthesia *middle,* there are typical EEG signs of the analgetic NLA phase (slower, stable alpha rhythm with activation in the delta/theta range). The modulations of the alpha activity point to a temporary insufficient cerebral suppression. The final phase *bottom* is determined by a considerable synchronization of the EEG at the transition of machine respiration to spontaneous breathing. Etiologically hypoxia has to be suspected due to still insufficient spontaneous breathing. Further wakening stimuli lead to a high-voltage alpha activity without significant low frequency.
Recording conditions: C_3-P_3 (*left*), C_z-A_1 (*right*), TC, 0.3 s; filter, 70 Hz; calibration, 50 μV = 7 mm; paper speed 30 mm/s; FFT, 30-s periods

Pat.: 79 yrs ♂

State of health: satisfactory

Lead: $C_Z - A_1$

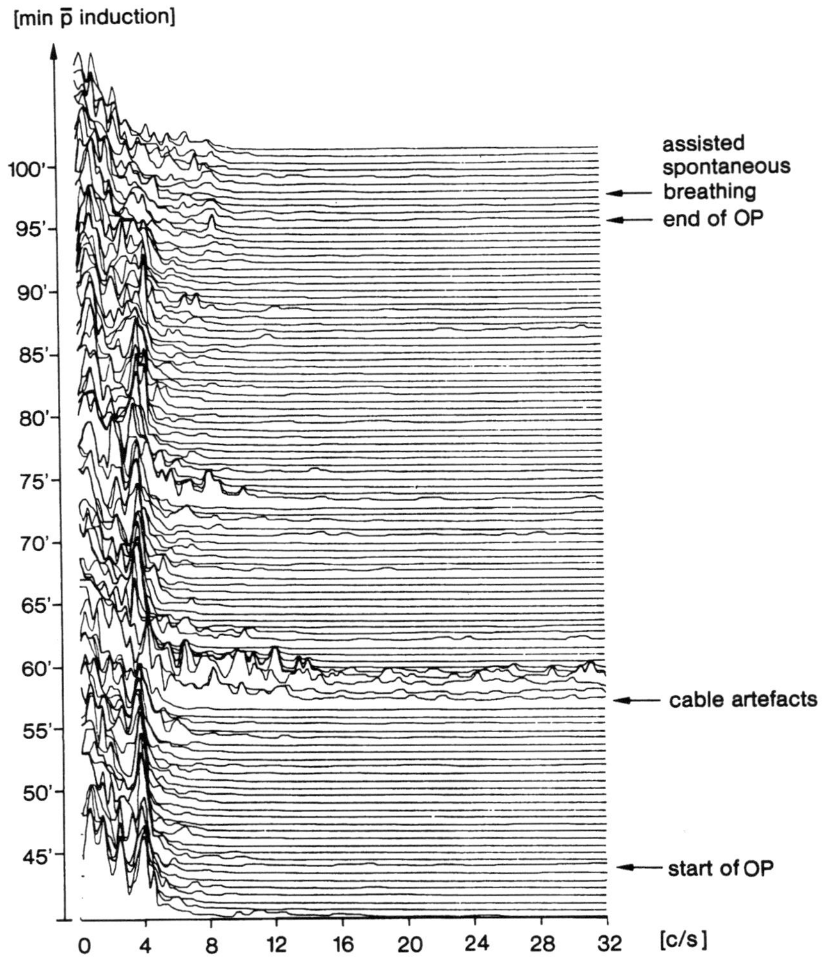

Fig. 5. EEG spectral analysis during barbiturate-induced neuroleptanalgesia for abdominal surgery. The EEG ist characterized by constant activation of the delta/theta bands with a stable 4-Hz peak. This pattern remains until after surgery at the beginning of spontaneous breathing. Contrary to the usual analgetic phase of the NLA, this patient demonstrates deep anesthesia, caused by fentanyl-droperidol, during the entire period of surgery.
Recording conditions: TC, 0.3 s; filter, 70 Hz; calibration, 50 μV = 7 mm; paper speed, 30 m/s; FFT, 30-s periods

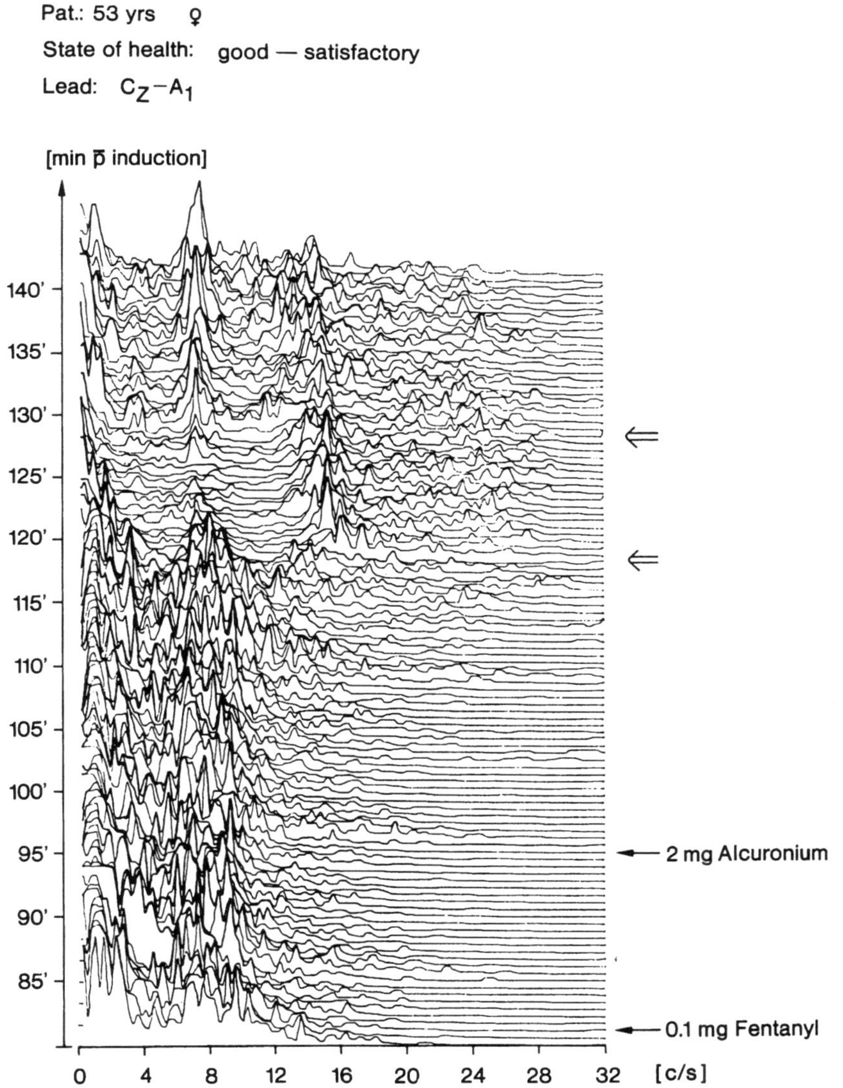

Fig. 6. Intraoperative arousal reaction, which is not clinically detectable because of the relaxation. From the steady state of barbiturate-induced neuroleptanalgesia, there is a sudden (⇐) almost complete disappearance of the typical NLA EEG activities with the development of moderate- to high-voltage beta rhythms. After additional anesthetic medication, the pattern shifts back to the analgetic phase of NLA. However, the continuing beta activity indicates that the previous stage of anesthesia is not achieved.
Recording conditions: TC, 0.3 s; filter, 70 Hz; calibration, 50 μV = 7 mm; paper speed, 30 mm/s; FFT, 30-s periods

Review of Other Findings on Intraoperative EEG Monitoring in Various Clinical Methods of Anesthesia. There are only a few reports on routine intraoperative EEG monitoring in the literature; they are mainly concerned with special problems. It is merely reported that the experimentally determined EEG stages of anesthesia are also expected to characterize the intraoperative pattern (PRIOR 1979). The advantages of intraoperative EEG monitoring are generally given as follows [4] (BRECHNER et al. 1962; SADOVE et al. 1967; PRIOR 1979):

- Maintenance of light anesthesia with curarization, without distress to the patient
- Avoidance of unduly deep stages of anesthesia
- Determination of any possible cerebral danger
- Forensic documentation of intact cerebral function during anesthesia
- Source of additional information in special operations or indications, e.g., open heart surgery or in patients with neurological symptoms

Previous experiments using automatic dose control by wave integration of certain filtered EEG frequencies have not as yet proven valuable [2]. Reasons for this are the lack of adequate technical equipment and the fact that it is not possible to detect potentially dangerous symptoms in such a simple manner. There are no reports on the routine use of intraoperative EEG monitoring without special indications.

General Anesthesia in Combination with Regional Anesthesia

The common anesthesiological technique of inserting an indwelling catheter into the peridural space has become more widely indicated in the treatment of severe postoperative pain, as described in Chap. B VI. These catheters are applied preoperatively. It would, thus, seem practical to use them intraoperatively, combining a classic high peridural regional anesthesia with a "light general anesthesia." Extinction of pain stimuli from the area of surgical incision and a "calming" of the sympathetic nervous system are the advantages of this method. This "calming," however, also leads to adverse effects due to vasodilatation of the area concerned. This, on the one hand, requires primary volume replacement, on the other hand it limits the compensatory mechanisms of the body against a lowered blood pressure during general anesthesia or against moderate blood losses during surgery. Advantages and disadvantages of the combined "general anesthesia and peridural anesthesia (PDA)," and their value for the individual patient, cannot as yet be determined in all aspects. The best choice of the adjunct hypnotic for anesthesia is still controversal. Two options that have so far been tested are now presented.

Thiopental-Enflurane-PDA

Thiopental-NLA-PDA

In 32 patients, PDA was combined with thiopental-enflurane anesthesia or thiopental-induced NLA. A PDA catheter was inserted and after fractionated application of the local anesthetic the effect of the regional anesthesia was observed. After sufficient analgesia up to the thoracic segments 12–10 was attained, the induction of the general anesthesia was performed according to standardized methods. As expected, the dose for induction did not differ from that in conventional anesthesia.

EEG Findings

Our Findings (16 patients 20–80 years of age, thiopental-enflurane-PDA, dose as required; 16 patients 20–80 years, thiopental-NLA-PDA, as required) (Tables 8, 9; Fig. 7). The intraoperative EEG is characterized by the typical changes in the particular general anesthesia. In the thiopental-enflurane group, there are the expected "lighter" stages of anesthesia.

Deeper levels of anesthesia are not recorded – sufficient anesthesia was found in 25% of cases, light anesthesia in 75%. In contrast to pure general anesthesia, the widely fluctuating depth of anesthesia is completely absent. Side effects of anesthesia – largely in the form of decreased blood pressure – were documented in 7 of 16 patients. In five cases (=31.2%), this led to adverse effects on electric activity.

A slightly different pattern of depth of anesthesia was seen in the combination with thiopental-induced NLA. Despite decreased consumption of anesthetic (Tables 8, 9), deep stages of anesthesia were registered in the EEG. Of the patients, 50% presented deep, or sufficiently deep, anesthesia, while 31.5% showed EEG signs of light anesthesia, and 19% widely fluctuating stages. In the resulting deeper level of anesthesia, only one of four cases, showing considerable decreases in blood pressure, demonstrated an altered cerebral function, as determined electroencephalographically.

Clinical Evaluation (Table 8). Both groups of patients presented similar states of health. The duration of surgery was definitely longer in the thiopental-NLA-PDA cases than in the groups with pure general anesthesia. In both forms of anesthesia, a significant saving of anesthetics and relaxing substances could be seen – 22% with enflurane, 40–50% with the drugs in NLA, and 20–25% with the muscle relaxants. Side effects, such as brief though considerable decreases in blood pressure, could not be avoided in 25% of the NLA combination cases nor in approximately 44% of the enflurane combination cases, despite simultaneously induced hypervolemia. Frequently, they had to be treated with vasoactive substances (especially if

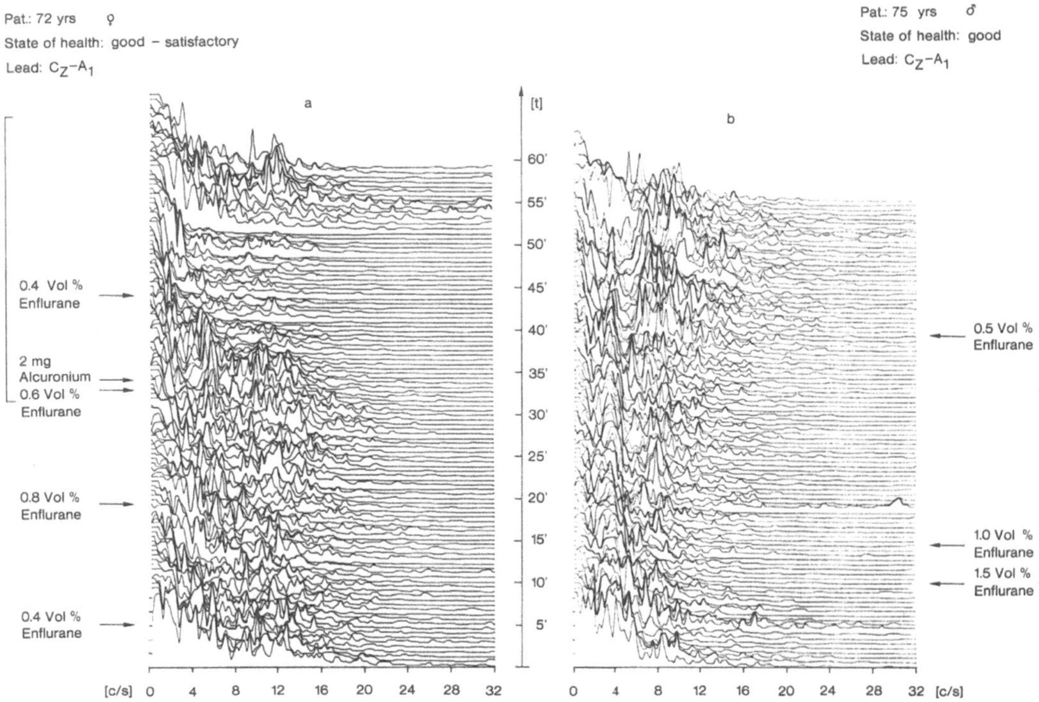

Fig. 7. a, b Comparison of the intraoperative EEG spectral analysis with thiopental-enflurane PDA **a** and thiopental-enflurane anesthesia. The combination of regional and general anesthesia **a** is characterized by activities in all four frequency bands as a light anesthetic stage, whereas the conventional general anesthesia **b** first has a moderate to deep anesthetic stage, which becomes lighter toward the end of surgery. Hypotensive episodes in both patients below the critical value of a mean arterial pressure of 70 mm Hg ([ ]) in light anesthesia indicate **a** a considerable impairment of the electric brain function. In contrast the same event in the stage of a deep anesthesia **b** leads to only mild functional changes, probably due to a decreased metabolism.
Recording conditions: TC, 0.3 s; filter, 70 Hz; calibration, 50 μV ≙ 7 mm; paper speed, 30 mm/s; FFT, 30-s periods

the blood pressure crisis proceeded to impaired cerebral function visible in the EEG). The rate of postsurgical machine respiration did not differ from that with conventional anesthetic methods.

Review of Other Findings. The reports of intraoperative EEG monitoring in "conventional anesthesia" deal only with general topics and do not mention special anesthesia. No recent studies are available regarding new forms of combination. SADOVE et al. (1967) report only that pure spinal anesthesia has no influence on the electric activity of the brain. Any changes are only as a result of adjunct psychopharmacological drugs.

Table 8. Anesthetic requirements and patient data with thiopental-enflurane anesthesia and thiopental NLA, combined with PDA ($\bar{x} \pm s\bar{x}$)

Drug	Thiopental-ethrane-PDA ($n = 16$)	Thiopental-NLA-PDA ($n = 16$)
Thiopental (mg/kg)	5.39 ± 1.20	5.06 ± 1.24
Succinylcholine (mg/kg)	1.39 ± 0.33	1.34 ± 0.28
Alcuronium (mg/kg/min)	10.7459×10^{-4} $\pm 5.0957 \times 10^{-4}$	10.617×10^{-4} $\pm 3.459 \times 10^{-4}$
Enflurane (l/kg/min)	7.9934×10^{-4} $\pm 1.4148 \times 10^{-4}$	– –
Fentanyl (mg/kg)	– –	5.0952×10^{-5} $\pm 2.4147 \times 10^{-5}$
Dehydrobenzperidol (DHB) (mg/kg/min)	–	58.7400×10^{-5} $\pm 22.5034 \times 10^{-5}$
Bupivacaine 0.375% (ml/kg/min)	1.6791×10^{-3} $\pm 1.1020 \times 10^{-3}$	1.6293×10^{-3} $\pm 0.7420 \times 10^{-3}$
Prilocaine 1% (ml/kg/min)	1.1858×10^{-3} $\pm 0.6305 \times 10^{-3}$	1.6867×10^{-3} $\pm 0.0283 \times 10^{-3}$
Age (years)	56.8 ± 17.2	59.9 ± 17.3
Artificial respiration in ICU (n, percent)	6 37.5	3 18.75
State of health (1–4)	2.21 ± 1.08	1.69 ± 0.66
Duration of surgery (min)	122 ± 41	192 ± 51

Table 9. Intraoperative EEG findings with thiopental-enflurane anesthesia ($n = 16$) or thiopental-NLA ($n = 16$) combined with PDA

Electroencephalographically defined depth of anesthesia	Thiopental-enflurane PDA		Thiopental-NLA PDA	
	n	Percent	n	Percent
Deep anesthesia delta/theta	0	0	3	18.75
Sufficient anesthesia delta/theta/alpha	4	25	5	31.25
Light anesthesia delta/theta alpha/beta-beta	12	75	5	31.25
Large fluctuations in depth of anesthesia delta/theta-alpha/theta	0	0	3	18.75
Impaired cerebral function due to exogenous factors	5	31.2	1	6.25
Side effects of anesthesia or surgery, e.g., blood pressure crisis	7	43.75	4	25

Interpretation of EEG Findings and Clinical Observations. The EEG is of special value in the newly developed forms of anesthesia described above. Through the lower anesthetic consumption in the PDA-enflurane combination, there are markedly lighter stages of anesthesia, as reliably indicated by the EEG. The greater influence of exogenous hazards on the cerebral function is also indicated in the EEG. The findings justify the EEG as an important monitoring device in this kind of anesthesia. It contributes well to the avoidance of cerebral damage in surgical or anesthesiological emergencies. NLA alone and in combination with PDA is of particular interest here. Despite significantly lower anesthetic consumption, the EEG shows anesthetic stages similar to conventional NLA. These findings demonstrate the positive effects of pain suppression by regional anesthesia in NLA. For optimization of the dose and determining the actual stage of anesthesia, the EEG is a valuable and indispensable aid.

Summary of Intraoperative EEG Monitoring

When using continuous EEG recording in the operating room, the following information can be expected:

- Evaluating parameters of the depth of anesthesia, avoidance of excessively light anesthesia
- Signs of possible cumulative drug effects undesired deepening of anesthesia
- Signs of exogenous impairment of cerebral function, with consecutive cerebral distress

The results show that – with a given classifiable initial EEG – the actual depth of anesthesia can be determined by the EEG in all the methods of anesthesia that have been presented here. The application of drugs causing specific EEG changes has to be considered when evaluating the depth of anesthesia. Cumulative effects of narcotics leading to inadvertent deepening of anesthesia are also visible in the EEG, and can thus be avoided. In any imbalance leading to impairment of cerebral function, the EEG is a sensitive parameter. Such phenomena provoked by decreased blood pressure or hypoxia could be observed in our own patients. CABRAL et al. also reported EEG signs in malignant hyperthermia [3], and PRIOR (1979) found seizure potentials that were caused intraoperatively by vascular anomalies. Hypoxia or anoxia due to respiratory conditions frequently cause a deceleration of frequency with a reduction in voltage, even before cardiovascular signs in the EEG. This should be taken as a most alarming signal in clinically constant anesthesia. In open heart surgery, the rate of postoperative cerebral damage has been markedly lowered alone as a result of EEG [1]. It is a striking yet consistent feature, corresponding to clinical observations, that our own findings reveal very light stages of anesthesia – purposely maintained, e.g., by the combination of regional anesthesia plus general anesthesia – to show a considerably greater intolerance of the brain to exogenous hazards, such as a decrease in blood pressure. In contrast, there seems to be less danger in deeper anesthesia. Intraoperative EEG monitoring thus represents another safety factor both for the patient and physician.

References

A. Textbooks and Summaries

Brechner VL, Walter RD, Dillon JB (1962) Practical electroencephalography for the anesthesiologist. Thomas, Springfield

Prior PF (1979) Monitoring cerebral function. Elsevier, Amsterdam

Sadove MS, Becka D, Gibbs FA (1967) Electroencephalography for anesthesiologists and surgeons. Pitman, London

B. Papers

1. Anonymous (1975) Brain damage ofter open-heart-surgery. Lancet II:399–400 (leading article)
2. Bickford RG (1950) Automatic electroencephalographic control of general anesthesia. Electroencephalogr Clin Neurophysiol 2:93–96
3. Cabral R, Prior PF, Scott DF, Brierley JB (1977) Reversible profound depression of cerebral electrical activity in hyperthermia. Electroencephalogr Clin Neurophysiol 42:697–701
4. Wiemers K, Puppel H (1960) Praktische Bedeutung der EEG-Registrierung im Operationssaal. In: Kreislaufmessungen: Vorträge des 2. Freiburger Kolloquiums über Kreislaufmessungen, München 1959, Bd 2, S 24–34

II. Electroencephalographic Monitoring During the Immediate Postoperative Phase

The immediate and early postoperative period is characterized by anesthetic hangover with the corresponding metabolic and general effects as a result of fluid, electrolyte, or acid-base imbalance. Metabolic changes also occur due to the operation and the strain exerted on the body by analgetically effective substances. The resulting cerebral metabolism, which is limited postoperatively, produces corresponding EEG changes. After extensive surgery, these changes are observed for up to 3 days, in a few cases even up to 2 weeks postoperation (SADOVE et al. 1967). EEG deviations when sustained longer, suggest cerebral damage due to surgery.

EEG Findings

Our Findings (95 patients 20–60 years of age, after moderate abdominal or gynecological surgery, average duration of operation 95 ± 38 min [6]) (Tables 1–3, Figs. 1, 2). Our own regularly recorded postoperative EEGs up to the 4th h postoperation showed qualitative changes compared with the individual initial readings, which differed with the various anesthesiological methods employed: In patients with an initial alpha EEG, there was a decrease in electric power and frequency of this alpha range, and an increase in activity in the delta band in the immediate postoperative period. Activities in the beta range were obtained in 30–50% of the patients. Variants without initial pathological values also showed postoperatively a slight modification of the initial activity, with increased activity in the delta/theta band. They were especially observed in connection with NLA.

Review of Other Findings. The existence of a postoperatively reduced electric cerebral activity for a period of hours to days has been reported ([11] SADOVE et al. 1967). GRABOW and PYKEL ([4] GRABOW 1981) proved that following surgery with the cessation of pain stimuli, an intensification of anesthesia occurs, which can be seen in the EEG and which is sustained for 2 h after halothane anesthesia and for up to 6 h after NLA. Patients undergoing halothane anesthesia return to their dominant initial frequency, while patients who undergo NLA still show a marked frequency deceleration.

In the quantitative interpretation of postoperative EEG findings, differences can be seen according to the method of anesthesia performed [5, 6, 8–10]. They are evident when the individual frequency bands are considered (e.g., average decrease in the alpha frequency after NLA is 1.2 Hz, after halothane 0.7 Hz), however, they become more noticeable when the important frequency ranges are evaluated together, e.g., by evaluating the alpha/delta ratio. Forty-five minutes after extubation, patients anesthetized with halothane have a ratio that is almost normal, with the corresponding clinical degree of alertness. Conversely, after NLA, patients show a ratio

Table 1. Postoperative EEG changes after barbiturate-induced halothane anesthesia ($n = 37$; observation period – 4 h postoperation)

Frequency range	Nature of EEG power change	Incidence (percent)	Degree of change
Alpha	a) ↓ DF b) ↑↓ DF	a) 83 b) 17	a) 1–2 Hz b) 1–2 Hz
Beta	↑	33	(14–16 Hz) 10–30 μV
Delta	↑	100	Varies individually
Theta	0	0	–

Abbreviation: DF, dominant frequency

Table 2. Postoperative EEG changes after barbiturate-induced NLA ($n = 35$; observation period – 4 h postoperation)

Frequency range	Nature of EEG power change[a]	Incidence (percent)	Degree of change
Alpha	↓ DF	33	1– 2 Hz
Beta	↑	17	(14–16 Hz) 10–30 μV
Delta	↑	100	Varies individually
Theta	0	0	0

[a] ↑ increase, ↓ decrease

Table 3. Postoperative EEG changes after barbiturate-induced fentanyle-N_2O/O_2 anesthesia ($n = 13$; observation period – 4 h postoperation)

Frequency range	Nature of EEG power change	Incidence (percent)	Degree of change
Alpha	a) ↓ DF b) ↑↓ DF	a) 33 b) 67	1– 2 Hz 1– 2 Hz
Beta	↑	40	(14–16 Hz) 10–30 μV
Delta	↑	100	Varies individually
Theta	0	0	–

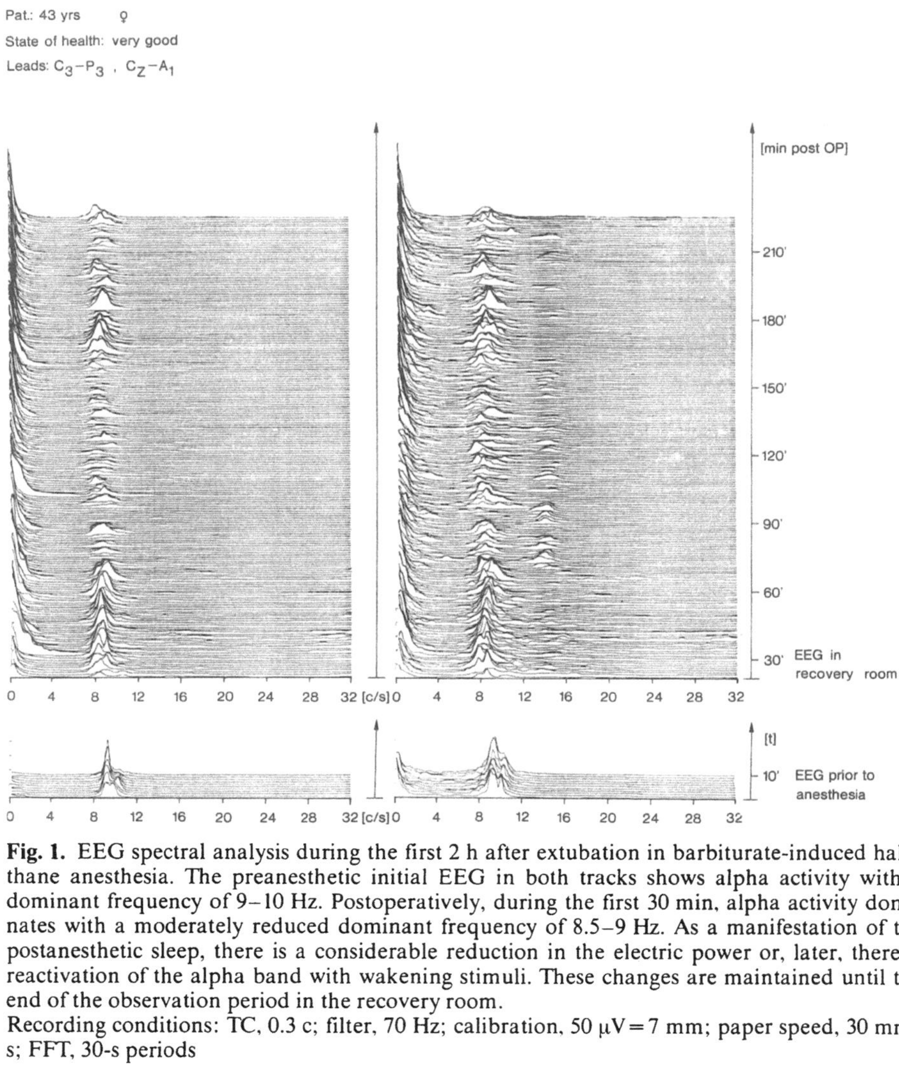

Fig. 1. EEG spectral analysis during the first 2 h after extubation in barbiturate-induced halothane anesthesia. The preanesthetic initial EEG in both tracks shows alpha activity with a dominant frequency of 9–10 Hz. Postoperatively, during the first 30 min, alpha activity dominates with a moderately reduced dominant frequency of 8.5–9 Hz. As a manifestation of the postanesthetic sleep, there is a considerable reduction in the electric power or, later, there is reactivation of the alpha band with wakening stimuli. These changes are maintained until the end of the observation period in the recovery room.
Recording conditions: TC, 0.3 c; filter, 70 Hz; calibration, 50 μV = 7 mm; paper speed, 30 mm/s; FFT, 30-s periods

that is markedly below normal, and there are no signs of a return to normal values until the end of the 4th h post operation.

Clinical Evaluation. Reliable criteria of vigilance in the postoperative period have not been determined. The clinical phase of postanesthetic sleep from immediately after the operation until the 4th h was determined in our own patients by the clinical degree of alertness and by psychological testing [5, 7]. A significant reduction in vigilance is present with corresponding postanesthetic sleep following halothane anesthesia and NLA. During the

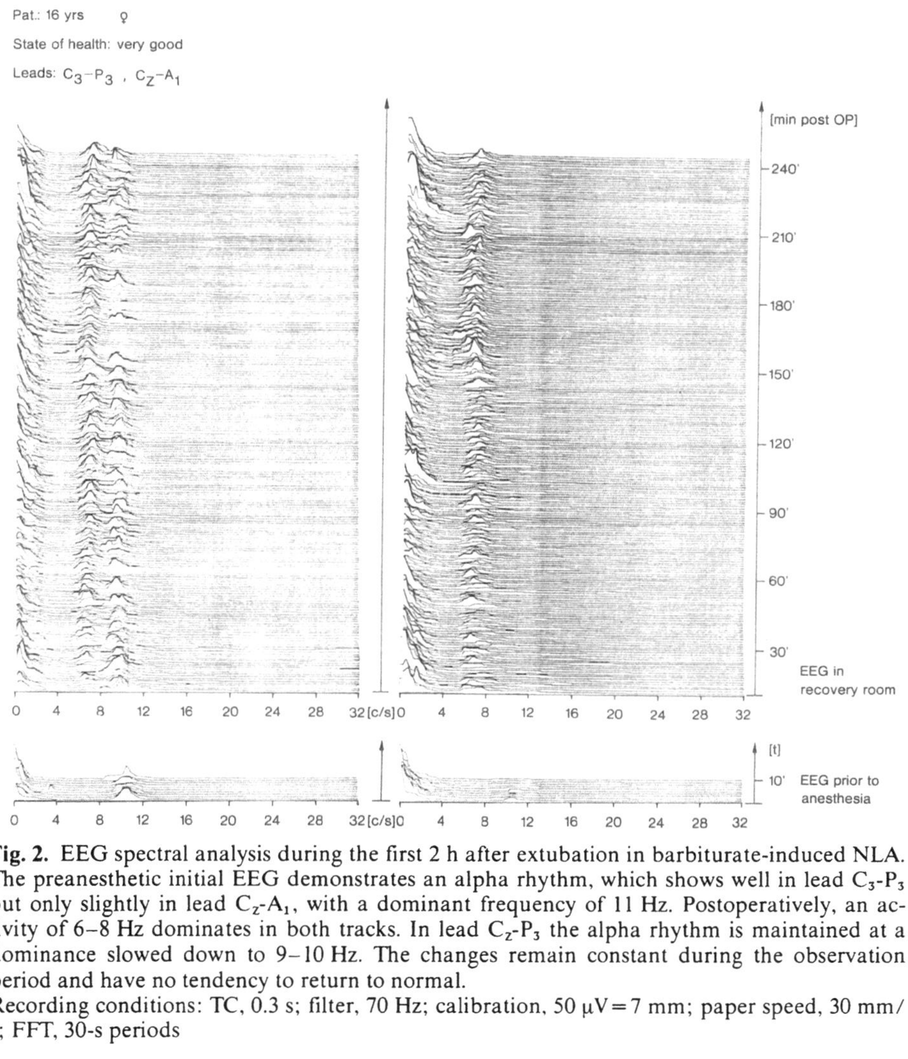

Fig. 2. EEG spectral analysis during the first 2 h after extubation in barbiturate-induced NLA. The preanesthetic initial EEG demonstrates an alpha rhythm, which shows well in lead C_3-P_3 but only slightly in lead C_Z-A_1, with a dominant frequency of 11 Hz. Postoperatively, an activity of 6–8 Hz dominates in both tracks. In lead C_Z-P_3 the alpha rhythm is maintained at a dominance slowed down to 9–10 Hz. The changes remain constant during the observation period and have no tendency to return to normal.
Recording conditions: TC, 0.3 s; filter, 70 Hz; calibration, 50 μV = 7 mm; paper speed, 30 mm/s; FFT, 30-s periods

first 4 h postoperation this shows only a slight tendency of returning to normal values. The reduced vigilance is significantly more marked following NLA according to all criteria. Similar results were obtained by GRABOW and PYKEL [3, 4]; they found a reduction in psychological skills up to 6 h following halothane anesthesia and NLA. DOENICKE et al. [1, 2] described somnolence persisting for up to 12 h after NLA.

Interpretation of EEG Changes and Clinical Findings. The EEG has proven to be a reliable indicator in evaluating postoperative functional reduction, which is clinically manifest as somnolence. The reduction in vigilance is visible in the EEG; it can be correlated with the clinical evaluation and the results of psychological testing. This reduction in vigilance is, however, more sensitive and indicate cerebral alterations over a longer period.

The significance of postoperative EEG recording is based on its determination of the overall postoperative situation and the detection of further changes due to postnarcotically applied drugs (analgetics). Additional medication during this period results in additive or potentiating effects. A postoperative reduction in vigilance and intellectual abilities, therefore, has to be expected. Normalization after halothane anesthesia occurs within 24 h; anesthetic hangovers due to NLA exceed the 24-h limit. These findings have to be taken into account when evaluating the recovery time after intraoperative incidents.

References

A. Textbooks and Summaries

Grabow L (1981) Hirnfunktionen unter dem Einfluß der Allgemeinen Anästhesie. Fischer, Stuttgart

Sadove MS, Becka D, Gibbs FA (1967) Electroencephalography for anesthesiologists and surgeons: Pitman, London

B. Papers

1. Doenicke A, Holle F, Kugler J, Kolle K (1964) The postnarcotic recovering time of the brain function with fentanyl and droperidol. 3. Congressus mundiales anaesthesiologicae, Sao Paulo, 1964, 20–26 Sep
2. Doenicke A, Kugler J, Schellenberger A, Gürtner T (1966) The use of the EEG to measure recovering time after intravenous anaesthesia. Br J Anaesth 38:580–590
3. Grabow L (1978) Zeitgleiche mehrfaktorielle Analyse der Hirnfunktion nach Halothan-Narkose. NLA und komb. Akupunktur-Analgesie. Anaesthesist 27:25–35
4. Grabow L, Pykel N (1980) Veränderungen der hirnelektrischen Aktivität unter dem Einfluß der allgemeinen Anästhesie. Anaesthesist 29:366–369
5. Gubernatis G (1980) EEG-Befunde im postoperativen Zeitraum unter Berücksichtigung der vorangegangenen Narkoseart. Inauguraldissertation, Medizinische Hochschule Hannover
6. Gubernatis G, Pichlmayr I, Künkel H (1977) Langfristige EEG-Frequenzanalyse nach barbituratinduzierter Halothan-Narkose und barbituratinduzierter Neuroleptanalgesie. Paper read at Zentraleuropäischer Anästhesie-Kongreß, Geneva, 13–17 Oct 1977
7. Lips U, Gubernatis G, Pichlmayr I, Tewes U, Alp O, Reichertz LP (1981) Ergebnisse einer psychologischen Beurteilungstestung und der EEG-Hintergrundaktivität im postoperativen Zeitraum nach 2 verschiedenen Narkosearten. Paper read at Zentraleuropäische Anästhesie-Kongreß, Berlin, 15–19 Sep 1981
8. Pichlmayr I, Gubernatis G, Luba A (1977) EEG-Verhalten in der Aufwachphase nach barbituratinduzierter Neuroleptanalgeise. Paper read at Zentraleuropäischer Anästhesie-Kongreß Geneva, 13–17 Oct 1977
9. Pichlmayr I, Gubernatis G, Luba A (1977) EEG-Verhalten in der Aufwachphase nach barbituratinduzierter Halothan-Narkose. Paper read at Zentraleuropäischer Anästhesie-Kongreß, Geneva, 13–17 Oct 1977
10. Pichlmayr I, Lips U, Gubernatis G (1978) Prä- und postoperative EEG-Überwachung am Beispiel einiger gebräuchlicher Narkosearten. Chir Forum Exp Klin Forsch pp 155–159
11. Trede M, Kubicki S, Just O (1959) Über EEG-Beobachtungen bei Herzoperationen mit extrakorporalem Kreislauf. Anaesthesist 8:72–82

III. Electroencephalographic Monitoring in Intensive Care

The methods of treatment that are now known as intensive care medicine were first developed around 1960. Since then, clinical experience and intensive research have enabled the development of more advanced medical and technical standards. It has also become possible to support or replace vital functions at least temporarily by technical-medical measures. The options or treatment have thus increased in number and have led to an expansion of surgical methods both with regard to the age and state of health of the patients and the extent of surgery. For these reasons, patients are nowadays frequently treated in intensive care units (ICU) after postoperative failure of one or more vital functions following major surgery or occasionally also after anesthesiological complications. These patients require long periods of treatment with very invasive therapeutic and monitoring methods. Apart from the drugs required for direct therapy, i.e., antibiotics or circulatory supporting drugs, larger amounts of anxiolytics, neuroleptics, sedatives, and hypnotics are needed for the mitigation of fear or stress caused by machine respiration; these drugs may potentiate the effects of analgetics and relaxants, which are also necessary. These phases of deep "long-term anesthesia," which may persist for some time, have considerable disadvantages: In the acute period of treatment, clinical evaluations of brain function are only possible to a limited extent; brief periods of awakening for orientation cannot be carried out due to the time of action and the accumulating effect of the substances used. After a successful phase of intensive care, there are further difficulties for the patient and physician: Psychomotor withdrawal symptoms after discontinuation of medication are followed by long periods of psychic instability, which impede the patient's recovery and reintegration into a social environment.

Drugs in intensive care are administered according to clinical demand; their purpose is to keep the patient from being aware of his desolate state. It remains to be clarified how far a decrease in sedation with extremely short-acting and rapidly metabolized substances may bring about a positive change for the post-therapeutic situation of the patient. Regular or continuous EEG checks during critical phases in a patient's development should provide information as to the depth of sedation and reflect the cerebral situation in emergencies brought about by the underlying disease. We also wanted to determine to what extent the EEG may represent a further aid in determining whether to introduce extensive therapy in irreversible cerebral failure [14].

EEG Findings

Our Findings in Patients Undergoing Postoperative Intensive Care (20 patients, 25–75 years of age; stay in ICU, 3–100 days; EEG recording, $n = 105$) (Figs. 1–3). The pattern of brain function was monitored by means of regular EEG recordings in 20 patients undergoing intensive care. The group of patients observed was heterogeneous; the patients had different medical histories and there were various reasons for their being admitted to the ICU: 16 patients were admitted with either renal failure (requiring dialysis) or primary septic complications with gradual involvement of all organs (a); in two patients a severe craniocerebral trauma (b) and in two other patients a cerebral hypoxia due to intraoperative hemorrhagic shock (c) preceded their admission.

Depending on the above-mentioned causes for admission to intensive care, the EEG recordings exhibit various primary findings and courses.

a) At the beginning of treatment, there are normal EEGs, which can be modified by therapeutic sedation. With insufficient medication, stress due to intensive care is shown by general desynchronization. The predominance of delta waves indicates deep sedation. The change toward a low-voltage, and subsequently to an isoelectric EEG without a change in the medication usually reflects general deterioration and proves prognostically unfavorable (Fig. 1). Effects of dialysis are evident in every case; they can be manifest both as a deterioration (cerebral edema after dialysis) and – more frequently – an improvement in cerebral function.

b) We observed a relatively small number of patients with craniocerebral trauma. At the beginning of treatment, the EEG shows bilateral differences and focal or even transient generalized seizure potentials. The transition to "level sections" indicates a deterioration of the clinical state (Fig. 2).

c) Intraoperative hypoxia gives rise to variably strong alterations in the EEG (delta/theta/alpha/beta) even in the immediate postoperative period. These alterations may resemble the initial findings in geriatric patients with arteriosclerotic changes. During brain-protective therapy with high barbiturate and cortisone doses, as is common today, the EEG findings are mainly characterized by barbiturate effects. Subsequent regeneration of cerebral function is evident in a gradual normalization of the frequency pattern (Fig. 3).

Review of Other Findings. The prime necessity in intensive care of replacing failing vital functions and the difficulties involved in including the EEG as a strictly "neurological specialty" in ICUs have thus for hindered wide use of the EEG in anesthesiological ICUs. Indication for an EEG in the literature reviewed is seen only in special emergencies (i.e., severe craniocerebral trauma, neurosurgical intensive care, previous cardiac arrests).

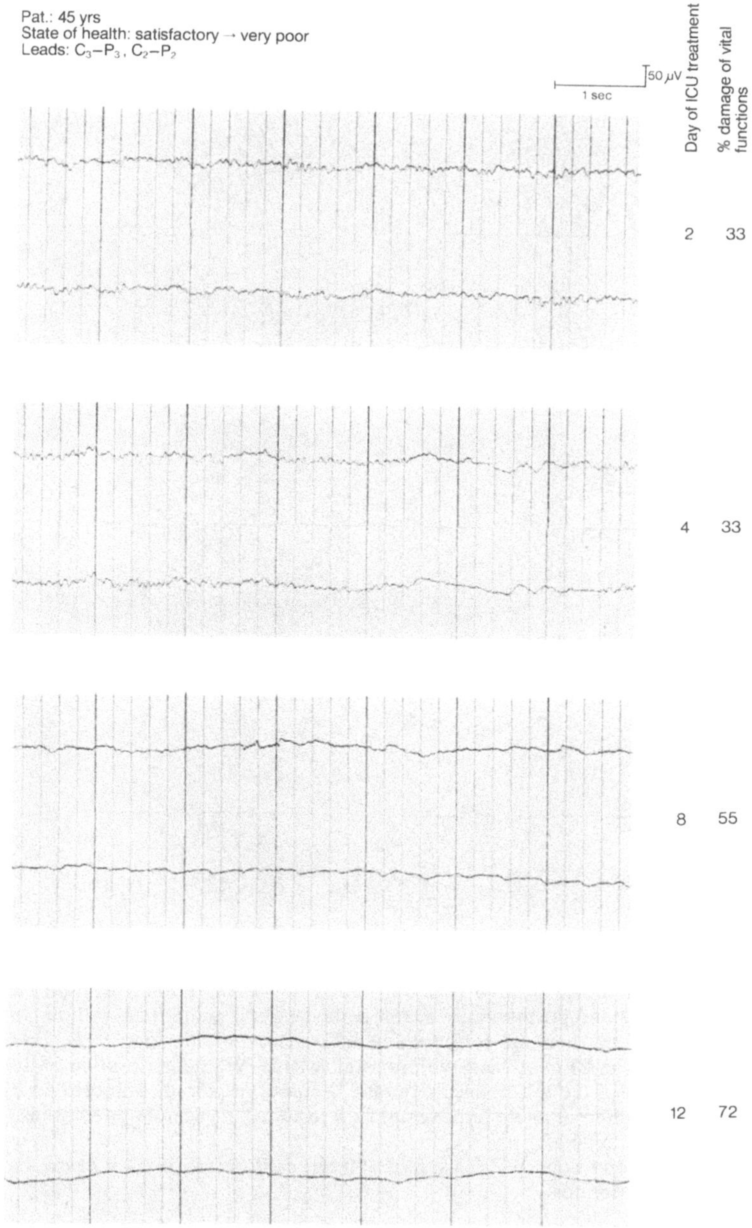

Fig. 1. EEG course of a 45-year-old patient on a respirator with acute postoperative renal failure and septicemia. At the beginning, a mixed alpha/beta activity with only few slow interspersed waves is seen; in the course of intensive care, there is an increase in frequency deceleration and lowering of the voltage, almost up to a completely isoelectric EEG with little activity. The changes demonstrate good concordance with the deterioration in the patient's condition. He died of intractable septic shock.
Recording conditions: TC, 0.3 s; filter, 70 Hz; calibration, 50 µV = 7 mm; paper speed, 30 mm/s; FFT, 30-s periods

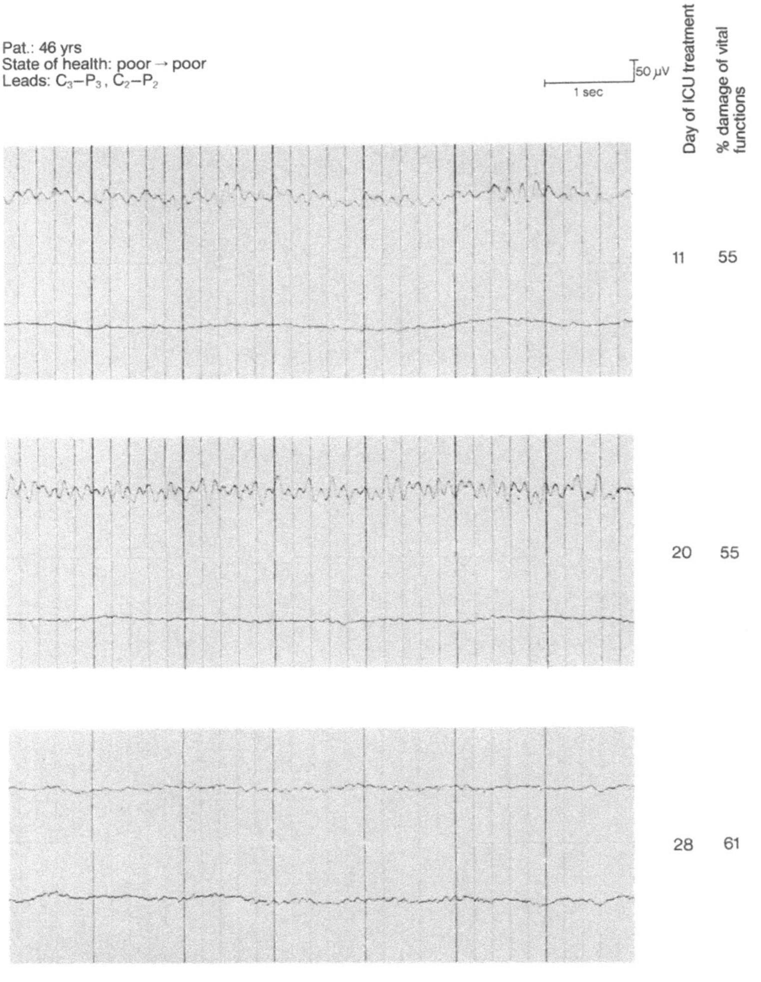

Fig. 2. EEG course of a 46-year old patient on a respirator with multiple injuries including head injuries and postoperative (orthopedic surgery) acute renal failure. As a manifestation of the head injury, there are considerable differences on both sides: *right,* almost isoelectric EEG; *left,* slow alpha EEG of 7 Hz, with normal voltage. With deterioration of the general condition, the function of the left hemisphere also becomes impaired, as indicated by a flattened EEG. The patient died of acute hypoxemia as a result of intractable pneumonia, which became fulminant with-in 24 h.
Recording conditions, TC, 0.3 s; filter, 70 Hz; calibration, 50 µV = 7 mm; paper speed, 30 mm/s; FFT, 30-s periods

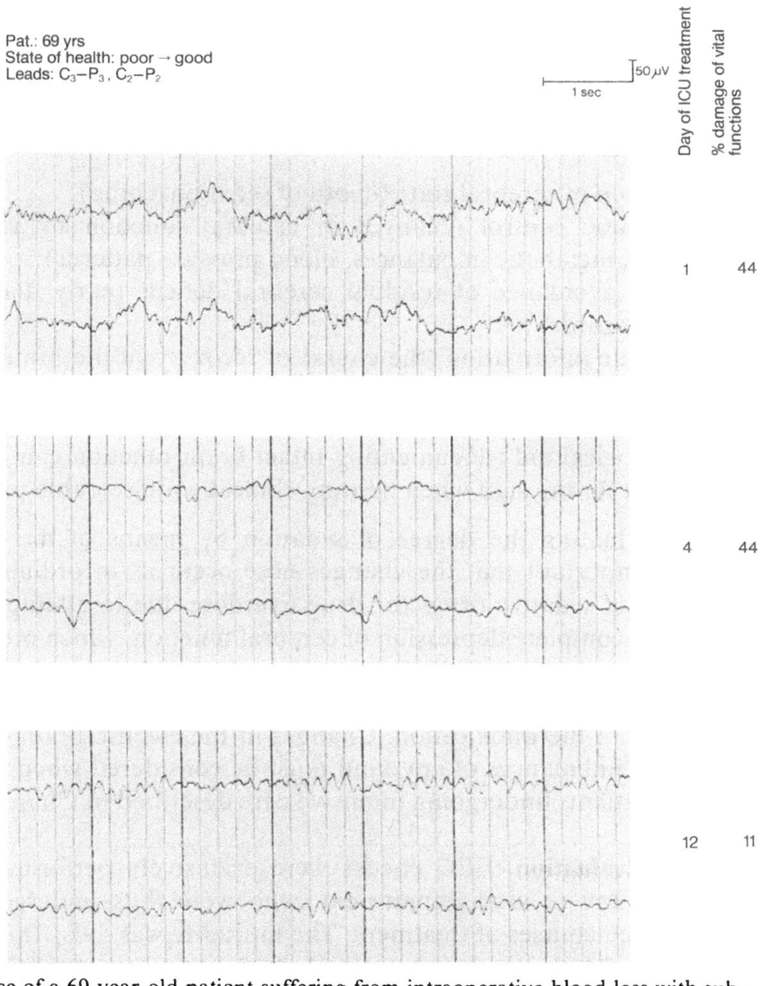

Fig. 3. EEG course of a 69-year-old patient suffering from intraoperative blood loss with subsequent cerebral ischemia for 15 min (clinically – dilated pupils not reacting to light). On admission to ICU, (1) the EEG shows irregular activities in all frequency bands. The course (2) is characterized by the effects of a high-dose therapy with thiopental; high-voltage delta activity points to a gradual diminishing of the barbiturate effects. Toward the end of intensive care (3), an almost normal alpha EEG is present, documenting the improvement compared with the initial situation. The patient was suitable for discharge to a general ward after 13 days without neurological impairment.
Recording conditions: TC, 0.3 s; filter, 70 Hz; calibration, 50 μV = 7 mm; paper speed, 30 mm/s; FFT, 30-s periods

Here, prognostic predictions of the cerebral course are only of limited value [2, 4, 9, 15]. Only PRIOR (1979) indicates the value of routine cerebral monitoring with the cerebral function monitor (see Chap. A.I.). Five important reasons for a general EEG monitoring of patients undergoing intensive care are mentioned:

1. Evaluation of cerebral state (depth of sedation, coma)
2. Therapeutic control (changes in cerebral function by respiratory influences, metabolic imbalances, blood pressure patterns)
3. Possible avoidance of residual cerebral deficits (early diagnosis of potential hazards)
4. Prognostic information (the course of recovery of the brain function following cerebral impairment may offer prognostic indications in favorable cases of the final neurological status)
5. Psychological aid (documenting intact brain function can be an encouragement for the staff when treating clinically unfavorable patients)

When evaluating the degree of sedation by means of the EEG it is especially important that the changes here occur as in ordinary anesthesia. Overdose of sedative drugs (or drug overdose due to attempted suicide) is evident as complete depression of cerebral function, which must not be confused with an irreversible failure of cerebral function. Activation of the EEG by external stimuli allows certain conclusions to be drawn regarding the extent of the intoxication. Changes in the electric brain pattern due to metabolic imbalances of any kind must be considered when evaluating the EEG of patients undergoing intensive care described in Chap. B VIII.

Clinical Evaluation. EEG checks were exclusively performed on patients with very severe vital impairment, who were thus undergoing long and complicated courses of treatment. The mortality was 59%. The patients were not responsive during the entire period of their stay in the ICU. Depending on the depth of sedation, they showed unspecific defensive reactions during treatment. In patients for whom treatment in the ICU was successful, a 1–2-week phase of gradual recovery of consciousness was observed.

Interpretation of EEG Findings and Clinical Observations. Alterations in cerebral function can be easily documented in the treatment of patients undergoing intensive care. The individual degree of sedation, the extent of cerebral damage, and the positive or negative sequelae of various therapeutic schemes are clearly evident and can be evaluated according to the resulting overall effects. Since the extent of the individual cerebral alteration frequently changes during a period of intensive care, prognostic predictions are not possible from the EEG alone, but may be made in conjunction with the general clinical situation. In patients with craniocerebral trauma, EEG findings can also be used for evaluating the clinical course.

The complex primary traumatic changes, however, can only be validly assessed by a neurophysiologically trained physician. Impairment of cerebral function by intraoperative incidents can be well documented in the recovery phase.

A special group of patients requiring intensive care is pregnant women with eclampsia. Eclampsia is the final stage of an untreated or therapeutically resistant toxemia of pregnancy. It usually develops between the 20th and 30th week of gestation. With an incidence of 0.6‰ of all births, eclampsia is a rare though dangerous and potentially fatal disease both for the mother and the fetus. Its drastic symptoms of tonoclonic generalized seizures are accompanied by multiple organic disorders, each of which can be fatal. Generalized edema, proteinuria, and excessive hypertension are, in addition to the seizures, the leading symptoms of the overt disease. They are caused by an arteriolar spasm, although the causative mechanism is still unknown. Therapeutic schemes, mainly for the therapy of cerebral seizures, are numerous at present – from the administration of magnesium ions and "massive sedation" to respiratory impairment, which may require artificial respiration. These methods involve a number of technical hazards, which are possibly dangerous for the patient. It was, therefore, suggested by Lips [13] that anticonvulsive therapy be carried out using antiepileptics and cerebrally dehydrating substances. The working hypothesis was that spasmophilia originates in cerebral volume overload following excessive increases in blood pressure caused by an impaired blood-brain barrier. This therapy, among others, was also monitored by regular EEG checks.

EEG Findings

Our Findings During Therapy for Eclampsia (11 patients with manifest eclampsia; one patient with first signs of postpartum epilepsy. Age 25 ± 4 years; period of treatment 3–17 days) (Figs. 4, 5). On admission, 11 patients presented an alpha EEG as the basic activity (7.5–12 Hz; 10–50 μV); one patient had a low-voltage EEG with a frequency of 7 Hz. The beta frequencies additionally present in all cases, may have been caused either by barbiturates or tranquilizers as part of the previous therapy. In 7 of 12 patients, the regular EEG checks revealed pathological wave patterns, even where there were no clinical signs of seizures during the treatment mentioned above. Here, the main EEG findings of eclampsia (six patients) was characterized by paroxysmal generalized theta eruptions and interspersed rhythmic theta groups, which were not related to drug effects or changes in vigilance (Fig. 4). The pathological pattern improved during drug therapy but could still be demonstrated to a mild degree in three patients even on the day of discharge. Normalization of cerebral function during therapy was seen in all patients as a slight frequency acceleration (1–2 Hz) and an increase in the voltage (by 10–20 μV) of the alpha activity.

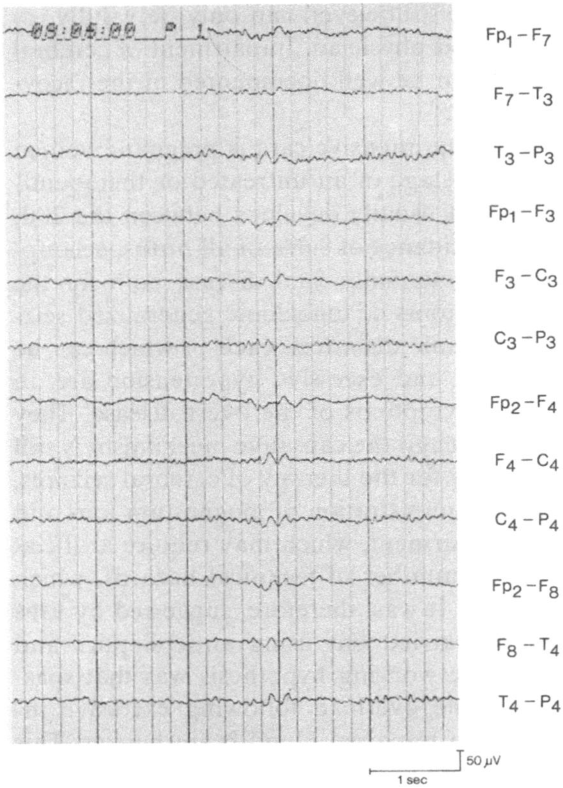

Fig. 4. EEG of a 32-year-old patient with overt eclampsia who presented no clinical symptoms (after phenytoin and dexamethason therapy). A general theta eruption is visible in the EEG tracking. This EEG pattern, which is typical for the CNS impairment resulting from eclampsia, did not recede completely during intensive care, despite clinical signs of remission.
Recording conditions: TC, 0.3 s; filter, 70 Hz; calibration, 50 µV = 7 mm; paper speed, 30 mm/s; FFT, 30-s periods

In one patient, the EEG examination revealed the first manifestation of a genuine epilepsy with spikes, multiple spikes, and waves (Fig. 5). In five patients, the EEG showed no abnormality.

Review of Other Findings. Due to the low incidence and the acute vital risk of onsetting eclampsia, EEG studies have been published only sporadically [6]. The findings reveal a very mixed picture. Both frequency deceleration with high-voltage beta activity [7] and dysrhythmias (up to 83% of the studied cases) [1, 3, 8, 10, 12, 18] without any possible classification of eclampsia have been described [5]. It is unclear whether such dysrhythmias occur only during the disease or whether they are indicative of a disposition to eclampsia [16, 17]. Although similarities in the eclampsia-epilepsy EEG are seen, especially during the convulsions [19, 20], differential diagnosis of the diseases is possible from the EEG [18]. One has to consider, however, that only in up to 64% of all grand mal epilepsies are unequivocal EEG changes detectable (Chap. B X). Clinical evaluation of the seizures, therefore, remains of significant value. The recovery time for EEG changes due to eclampsia is given as 3–8 weeks [10, 18].

Clinical Evaluation. During the treatment of eclampsia with the scheme mentioned above, a remission of the central nervous manifestations (hyperreflexia, seizures) was achieved in all patients. An advantage of this method is the preservation of the patient's cooperation and sense of well-being without the need for simultaneous artificial respiration. Mild sedation

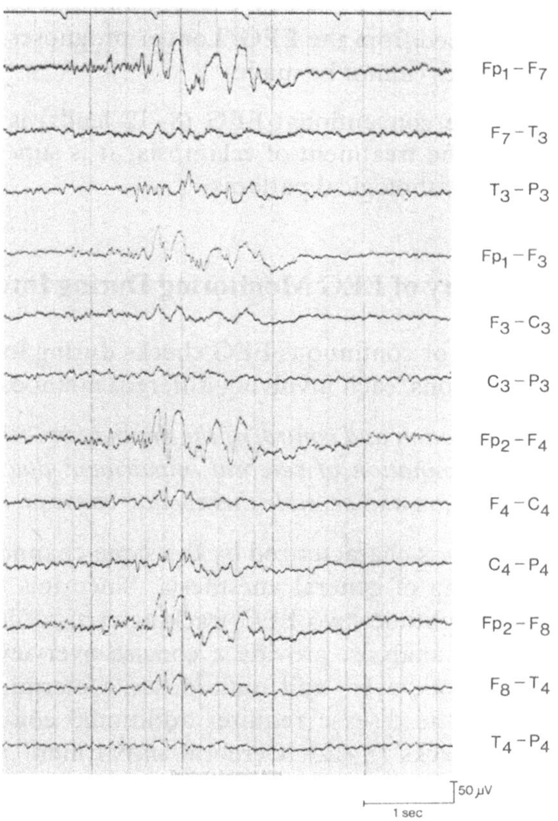

Fig. 5. EEG of a 21-year-old patient suffering postpartually from a tonoclonic seizure. The initial diagnosis of a postpartal eclampsia had to be revised to postpartal manifestation of epilepsia on the grounds of the EEG findings. The EEG shows the spike waves and polyspikes proving epilepsia in overt seizures. After therapy with phenytoin and dexamethazone, the patient was free of symptoms, the EEG returned to normal. The patient did not undergo the recommended neurological consultation upon discharge.
Recording conditions: TC, 0.3 s; filter, 70 Hz; calibration, 50 μV = 7 mm; paper speed, 30 mm/s; FFT, 30-s periods

at night helped to shield the patient from an uncomfortable mental disposition, which cannot be avoided completely in intensive care. The therapy of the remaining organic disorders was performed according to the appropriate general principles.

Interpretation of EEG Findings and Clinical Observations. Both in our own findings as well as in the reference material, the following advantages of EEG monitoring during the management of eclampsia are seen:

- The differential diagnosis of eclampsia-epilepsy can be made with the limitations mentioned above
- The extent of the cerebral impairment due to eclampsia is visible and allows guidelines for subsequent therapy
- An acute paroxysm of seizures and the efficiency of therapy can be determined from the EEG. Longer prognoses as to the occurrence of further seizures cannot be made

Only the conventional EEG (8–12 leads) is suitable for EEG monitoring during the treatment of eclampsia; it is superior to spectral analysis in detecting pathological patterns.

Summary of EEG Monitoring During Intensive Care

Regular or continuous EEG checks during intensive care are performed for two reasons, each involving different methods of recording:

- *Evaluation and control of the depth of sedation*
- *Determination of cerebral impairment due to the underlying pathological condition and possible therapeutic consequences*

Sedation is characterized by the same changes in cerebral function as at the beginning of general anesthesia. Since this is a general shift in frequency and amplitude, two EEG trackings are sufficient, which, in the manner of spectral analysis, provide a concise overview of the sedation and help to avoid both over- and underdoses. Determination of cerebral impairment due to the disease requires additional conventional EEG recording with more tracks (8–12). Cerebral impairment may occur locally (focally) or may be generalized, thus requiring the detection and evaluation of the appropriate graphoelements. Under these conditions, conclusions as to further cerebral deterioration can be made with respect to functional changes due to drugs.

References

A. Textbooks and Summaries

Prior PF (1979) Monitoring cerebral function. Elsevier, Amsterdam

B. Papers

1. Aresin L (1961) Elektroenzephalographische Untersuchungen bei Schwangerschaftstoxikosen. Z Geburtshilfe Gyn 157:235–240
2. Binnie CD (1975) The EEG in intensive care: Interpretation. J Electrophysiol Techn 1:5–15
3. Brehm R (1970) Zur Therapie der Spätgestosen. Z Geburtshilfe Gyn 173:31–41
4. Bushart W, Rittmeyer P (1966) Die Bedeutung des EEG-Befundes im Rahmen der Intensivbehandlung. Anaesthesiol Wiederbeleb 17:71–82
5. Douglass LH, Morrison JH, Oster R (1953) Cerebral dysrhythmia and acute toxemia of pregnancy. Obstet Gynecol 1:287–288
6. Gestaut H (1970) Funktionelle Epilepsien. Boehringer-Sohn, Ingelheim
7. Gibbs FA, Reid DE (1942) The electroencephalogram in pregnancy. Am J Obstet Gynecol 44:672–675
8. Gouskos AD, Aravantinos D, Chalkiadakis J, Kranidiotis PT (1974) The significance of electroencephalogram in the toxemic patient. In: Rippert TH, Rippmann ET (eds) EPH-gestosis. Huber, Bern, p 231–233
9. Hoffmann M, Scherzer E (1973) Enzephalographische und klinische Probleme nach passagerem, durch intrathorakale Herzmassage behobenem Herz- und Atemstillstand. Wien Z Nervenheilkd 31:291–304
10. Kolstad P (1961) The practical value of electroencephalography in pre-eclampsia and eclampsia. Acta Obstet Gynecol Scand 40:127–138
11. Kurtz D, Naquet R (1977) Etude électroencéphalographique des encéphalopathies functionelles dysmetaboliques et des comas au cours de l'alimentation parentérale. Ann Anesthesiol Fr 12:997–1006
12. Ledermaier O, Niedermeyer E (1956) Posteklamptische Epilepsie. Geburtshilfe Frauenheilkd 16:679–685
13. Lips U (1982) Medikamentöse Therapie der Eklampsie. Vorstellung einer neuen Methode zur Lösung eines geburtshilflich-anästhesiologischen Gemeinschaftsproblemes auf der Intensivpflegeeinheit. Habilitationsschrift, Medizinische Hochschule Hannover
14. Lorenz R (1969) Kriterien der Hirntätigkeit in lebensbedrohten Zuständen – ein Beitrag zur Frage des zentralen Todes. Acta Neurochir (Wien) 20:309–329
15. Lorenzoni E (1975) Das EEG im posttraumatischen Koma. Neurol Psychiatr 43:155–191
16. Rosenbaum M, Maltby GL (1943) Cerebral dysrhythmia in relation to eclampsia. Arch Neurol Psychiatr 49:204–213
17. Schroeder C (1951) Gelingt es mit Hilfe der Elektroenzephalographie, den Ausbruch von eklamptischen Krämpfen bei Präeklampsie vorherzusagen? Med Welt 20:306–309
18. Schroeder C (1955) Zur Differentialdiagnose: Eklampsie-Epilepsie. Arch Gynaecol 186:171–175
19. Torgård E, Brody S, Dhuner KG (1965) EEG in eclampsia. Abstracts of 6th international congress of electroencephalography and clinical neurophysiology. Vienna, 5–10 Sep 1965. pp 69–75
20. Whitacre FE, Loeb WM, Chin H (1947) A contribution to the study of eclampsia. J Am Med Assoc 133:445–449

Final Remarks

The EEG measures recognized parameters of cerebral function in neurophysiology. It is sensitive and can provide acute indications of both general (sleep EEG, pharmacological EEG) and specific hazardous cerebral changes. Its routine application in surgery with extracorporal circulation has drastically decreased the rate of postoperative cerebral disorders. Considerable progress has been achieved in anesthesiology, surgery, resuscitation and intensive care and this has led to extension of the surgical indications with regard to age, previous medical history, and duration and severity of the procedure. Impairments of cerebral circulation and oxygen supply due to anesthesiological or surgical emergencies become manifest during recovery as multiple disturbed functions. Thus, the EEG has become increasingly important as a monitoring aid of cerebral function. For the anesthesiologist in particular, the following advantages of EEG monitorings are given in the reference material:

- The possibility of maintaining an even level of anesthesia
- Early detection of inadequate cerebral metabolism
- Information about the cerebral bioavailability of newly introduced centrally acting substances
- Knowledge of the extent of possible cerebral damage as a result of unforeseen anesthesiological or surgical events as well as monitoring of further developments

Our experience with the practical application of EEG monitoring in routine anesthesia and intensive care has revealed: Anesthesia, when competently performed, is at present determined by the behavior of circulatory parameters in accordance with general anesthesiological and clinical criteria. Depending on the cardiovascular side effects of the drugs, large differences in the depth and functional stage of anesthesia are evident in the EEG. Nowadays, combination anesthesia using muscle relaxants or additional conduction anesthetics can provide analgesia as well as clear and unrestricted conditions for the surgeon not only with deep but also with moderate, light, and extremely light stages of anesthesia. The ideal depth of anesthesia for patients undergoing surgery, both generally and in particular cases, has yet to be established.

- Under very light anesthesia, cerebral emergencies cause severe, and thus more clearly visible, EEG changes. In cerebral insufficiencies, where these can be well determined, medium stages of anesthesia appear to exert a protective effect on the brain due to a measurable decrease in metabolism.

- Patients with a "healthy initial EEG" react under conditions of a cerebrally deficient circulation more quickly and strongly to functional changes than do patients with an initially pathological EEG. In the latter, a decrease in the cerebral rate of metabolism can be assumed.
- The use of regular EEG checks on patients in intensive care can allow important therapeutic and prognostic conclusions to be made, provided the limits of interpretation are not exeeded.
- Conclusions arrived at from theoretical considerations and reports in the reference material can be confirmed clinically by the EEG.
- For the anesthesiologist, EEG monitoring in daily practice can be carried out with little additional consumption of time.
- EEG assessment in the context of clinical findings requires the appropriate basic neurophysiological knowledge of the anesthesiologist in the hospital. For anesthetic monitoring, this is possible without long and specialized EEG training. A brief period of instruction in an EEG department would nevertheless be desirable.

Continuous EEG monitoring of anesthesiological procedures is recommended. The resulting experience can become basis for considerable improvements in anesthesiological therapy. Continuous EEG monitoring may disclose early cerebral deficiency symptoms, thus preventing transient or permanent cerebral damage. For the use fo EEG monitoring in special fields the following should be aimed toward:

- On the one hand, there should be intensive study of the effective spectra of the drugs used with regard to the CNS and its functional areas; on the other hand studies of methods of EEG registration and assessment should be carried out as part of a specialist's training.
- Recognition of a one- to two-track EEG recording solely as a general cerebral monitoring aid by the specialties of neurology and neurophysiology; it should be appreciated that a diagnostic EEG of an appropriately equipped neurophysiological institute cannot be replaced by one- or two-track EEG monitoring.
- Simplification of the EEG registration in the near future, since at present it is fairly expensive technically and requires a great deal of space. Also desirable is a series of small, resistant, and cheap recording devices; they should be able to register the conventional EEG curve and document the corresponding spectral analytic patterns.

The EEG is not only a monitoring instrument but also offers a possibility of checking the individual effects of a specific therapy. For these reasons, our book has been designed for the anesthesiologist in daily practice. It is meant to convey the fundamentals of general EEG monitoring, which is desirable in many anesthesiological fields.

Subject Index

MIX
Papier aus verantwortungsvollen Quellen
Paper from responsible sources
FSC® C105338

If you have any concerns about our products,
you can contact us on
ProductSafety@springernature.com

In case Publisher is established outside the EU,
the EU authorized representative is:
Springer Nature Customer Service Center GmbH
Europaplatz 3, 69115 Heidelberg, Germany

Printed by Libri Plureos GmbH
in Hamburg, Germany